Genetics for Dog Breeders

GENETICS FOR DOG BREEDERS

Frederick B. Hutt
Cornell University

W. H. FREEMAN AND COMPANY
SAN FRANCISCO

Sponsoring Editor: Gunder Hefta
Project Editor: Pearl C. Vapnek
Copy Editor: Linda Purrington
Designer: Michael Clane Graves
Production Coordinator: Fran Mitchell
Illustration Coordinator: Cheryl Nufer
Compositor: Typesetting Services of California
Printer and Binder: The Maple-Vail Book Manufacturing Group

Library of Congress Cataloging in Publication Data

Hutt, Frederick B 1897–
Genetics for dog breeders.

Bibliography: p.
Includes indexes.
1. Dog breeding. 2. Dogs—Genetics. I. Title.
SF427. 636.7'08'21 79-15169
ISBN 0-7167-1069-2

Printed in the United States of America

9 8 7 6 5 4 3 2 1

To Alice Jean B. Hutt

Contents

Preface

My veterinary students (who, during twenty years, taught me much) seemed to learn genetic principles that were illustrated by examples in dogs more easily than those with examples from chickens or fruit flies. Accordingly, I had to become knowledgeable about heredity in dogs.

As word got around, I had the pleasure of talking to kennel clubs and of hearing at first hand the problems of dog breeders. After finding most dog breeders desirous of knowing more about genetics than the 3 : 1 ratios that they had learned in high school, it seemed to me that a book on the subject, written specially for dog breeders, might be helpful.

Most breeders are concerned about the hereditary defects that have appeared in many a breed and many a kennel. In Chapters 7 to 12 I review our knowledge of many of those defects. Descriptions of them deal more with the clinical symptoms that the dog owner sees than with the underlying abnormalities that cause the outward and visible signs. Readers wishing to know more about the pathology involved should obtain details from their veterinarians or from the pertinent references cited.

In discussing some of these defects, I have tossed in a few ideas that may be regarded as rank heresy by some dog breeders, or by some veterinarians, or by both. This is done, not to irritate either group, but with the hope that the heresy might stimulate some thought and, eventually, some further research.

A special effort has been made to show the kinds of pedigree charts that are really informative and to describe the kinds of experimental matings desirable to prove conclusively the mode of inheritance. It is to be hoped that future studies of the dog's genes will be based more on experimental matings than on case reports, common ancestors, and a designation as "familial." When the mode of inheritance is known, it becomes easier to eliminate unwanted genes.

In Chapter 14, the reader will find some suggestions for breeding better dogs. They are based on principles and practices that I have found to be effective in breeding better chickens. Admittedly, we can make progeny-tests more easily in a prolific species like the domestic fowl than in dogs, but I have suggested ways in which co-operating dog breeders might overcome the handicap of somewhat slower reproduction in dogs, and thus make use of progeny-tests. These are not likely to produce any more or better champions than we have now, but they might result in less hip dysplasia and less of numerous other defects for which the genetic basis seems to be polygenic and complex.

I am grateful to the many colleagues and breeders who have generously supplied photographs and to the publishers who have permitted reproduction of some of their illustrations. Rather than a list of these here, a suitable credit line is given with each figure. If an illustration has been reproduced without change, that credit reads *from*; but when some modification has been made by my publisher's artists, it reads *after*. Walter Chimel, of Gaines Professional Services, has been helpful in rounding up illustrations.

I thank those who read parts of the manuscript to ensure that I did not wander too far astray. These include Drs. W. Jean Dodds (bleeding diseases in Chapters 4 and 10), R. C. Riis (Chapter 11), R. K. Cole and J. R. Smyth, Jr. (Chapter 14). Dr. R. D. Crawford reviewed the entire manuscript and made many helpful suggestions. Dr. W. T. Federer kindly contributed to Table 6-5, as there noted.

I must also thank Diane Bondioli for her efficient help, as editor and typist, in the preparation of the manuscript. Assurance in advance that her skills would be available made it easier for me to undertake this book than it would otherwise have been.

May 1979 F. B. Hutt

Genetics for Dog Breeders

Introduction

Just as I was starting to write this book, there came to my notice an article advising hopeful authors of prospective best-sellers that the best way to convert a browser into a buyer is to start off with an irresistibly intriguing opening sentence. The first example given was "'Hell!' said the Duchess." As any route from such a start to genetic control of hip dysplasia in dogs would be very circuitous, it seems better to dispense with profanity and stick to brass tacks.

Since the rediscovery of Mendel's laws in 1900, the science of genetics has grown, expanded, and branched into fields of interest so diverse that the specialist in one of them may know little or nothing about the advances in another. For example, the biochemist probing the processes by which the deoxyribonucleic acid (DNA) of the gene gets its message to those elements of the cell from which tissues are built is not likely to spend much time on the theories of the population geneticist about the preservation, spread, or extinction of genetic variations. Cytogeneticists specialize in the aberrant behavior of chromosomes. In the world's agricultural colleges, there is a whole army of geneticists breeding plants and animals to yield more food for the ever-increasing population of this planet. That population owes much to the plant breeders who in the last 60 years have bred grains, vegetables, and fruits for greater yields, for resistance to disease, and for adaptation to environments once thought to be too severe.

For any dog breeder interested in these fascinating branches of genetics, or other specialties, this is not the right book. There are plenty of others. This one is written chiefly to tell the dog breeder what principles of genetics can be useful in the effort to breed dogs conforming to his ideals and carrying a minimum of the "genetic junk" that plagues all species, whether wild or domesticated, whether mouse or man, and all breeds, whether St. Bernard or Yorkshire Terrier.

That term, "genetic junk," is just a useful umbrella under which to group undesirable hereditary deviations from normality that affect form or function. Many of them will be described in the chapters to follow. They result mostly from changes in the genes, but a few arise from aberrant behavior of chromosomes. (It is assumed that, in this day and age, every dog breeder knows that a gene is a unit of inheritance, too small to be seen under a microscope, and that the chromosomes, which carry the genes, can be seen as tiny threads under the microscope if the cells containing them were suitably stained when at the best stages for study.) These changes are known as *mutations*. That term can be applied either to the invisible change in the gene or to the visible variation that it causes.

Mutations occur spontaneously in animals and plants, and for some of them the frequencies with which they do so have been calculated by geneticists more venturesome than this author. Animals showing mutations were commonly known as "sports" in the years before genetics became a science.

In this book, we shall read mostly about mutations that are undesirable, but, before doing so, we must recognize that not all mutations are bad. Many of them that have occurred since breeders began to improve plants and animals are good and desirable. They are the ones that cause the variation that enables the breeders to select for whatever objectives they desire. For example, breeders of wheat in North America have accumulated in modern varieties mutant genes that induce greater yields per acre, wheats of superior milling quality, some with shorter straws, many with stiffer straw (to prevent lodging), varieties resistant to stem rust and to other diseases, and even fast-maturing varieties that have extended the wheat acreage northward far beyond the limits once imposed on the growing of wheat by early frosts. The cultivated fruits and vegetables on which we depend so much for our survival have all been improved by constant selection for better yields, better quality, or for both. Similar selection in domestic animals for better conformation, faster growth, and more efficient conversion of grain and grass to meat has been highly effective. Without continuous variation and selection, the modern efficient domestic animals could not have been bred. Mutations, most of them with only slight effects, have made that selection possible.

In the dog world, most breeders would classify as desirable mutations those which have resulted in the differentiation of breeds according to the differing ideals and objectives of dog breeders in different parts of the world. Breeds have been selected for long legs and short ones, for long faces and short ones, for large bodies and small ones, and for almost as many colors and patterns as in Joseph's coat. They have been selected not only for form but also, as is attested by many breeds, for function. Just how much of the latter is attributable to inheritance and how much to training is a question best decided by breeders more familiar than this author with wolf-hounds, deer-hounds, sheepdogs, bird-dogs of various kinds, fox-hounds, otter-hounds, and Dalmatians. The point is that selection has enhanced the skills of many breeds and that the selection has been mostly a matter of accumulating variations (i.e., desirable mutant genes) most conducive to the objective sought by the breeder.

Before going on to consider how mutations show up in the kennel, and what we can do about them, let us pay a silent tribute to the memory of the man who started it all—Gregor Mendel (Figure 1). His famous experiments with garden peas were made in the 1860s in the city now known as Brno, in Czechoslovakia. Mendel was a monk in the local monastery and a teacher of science in the high school.

He studied contrasting paired characteristics, such as tall and short plants, seeds wrinkled or smooth, and other variations. When parents differing in some such characteristic were crossed, one member of each pair showed up in all the first generations, while its contrasting partner did not. The *dominant* type prevented its *recessive* partner from appearing. In the second generation, both kinds occurred, and in the approximate ratio of 3 dominant : 1 recessive. Further details are unwarranted in this book, but all breeders of dogs and the scribes who write about Mendelian ratios should know the origin of that term.

Mendel was the first geneticist, but he did not know that word. The term *genetics* was not used until 1906, a whole 40 years after Mendel's findings were published, when it was proposed by William Bateson for the then new science dealing with heredity and variation.

Let us now consider how mutations show up in the kennel and what we can do about them.

FIGURE 1

Gregor Johann Mendel, 1822–1884. A picture taken at the time of his research. [Courtesy of Mendel Museum, Brno, Czechoslovakia.]

Part I

Principles of Genetics

Exemplified by Hereditary Variations in Dogs

1 Single Genes and Simple Genetics

Simple Recessive Mutations

The kind of mutation most easily recognized is one that is evident at birth (or soon thereafter) and shows such marked deviation from normal, or from the type expected, that the abnormality is seen at a glance. Let us consider one of these in order to acquaint ourselves with some commonly used terms in the geneticist's vocabulary.

A nice example was provided by the "bird-tongue" mutant studied by the author a few years ago (Hutt and de Lahunta, 1971). That was the breeder's name for it, and it was a good one, because all the affected puppies were found to have tongues narrower than normal, especially toward the tip (Figure 1-1). However, since no geneticist could ever hope to impress his colleagues with a nice, clear term like that, the name used in the published report was "lethal glossopharyngeal defect." That was all right too, but it will be simpler for our purposes to label the condition *bird-tongue,* as it was in the beginning.

Affected pups appeared at birth to be normal, but they made no attempt to suckle, and, in spite of all attempts to teach them to suckle, they died from starvation within a few days of birth. The owner, who had been a nurse as well as an experienced dog breeder, found by appropriate tests that the bird-tongue puppies did not swallow and could not be taught to do so.

My veterinary associate in this study could find no abnormality in the nerves or muscles of the bird-tongue puppies that could prevent swallowing. One might suppose that the inward folding of the margins of the tongues could interfere with the normal suckling process, but, as the pups made no attempt whatever to suckle, other associated abnormalities (such as faulty innervation of essential muscles) must have been involved.

The defect is clearly caused by a *recessive* gene in the *homozygous* state. Genes occur in pairs, as do the chromosomes that carry them. The dog has 39 such pairs of chromosomes, each member (in all pairs but one) appearing under the microscope to be the exact counterpart in size and shape of its partner. Similarly every gene in one chromosome normally has a partner at the same *locus* (position) in its matching

(a)

(b)

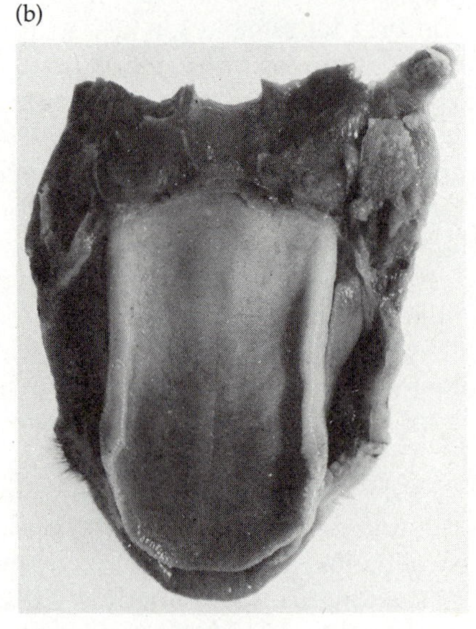

(c)

FIGURE 1-1
Tongues from (a,b) two bird-tongue puppies and (c) a normal one of comparable age. age. Note the inward folding of the margins of the outer halves of the affected tongues. [From Hutt and de Lahunta, 1971.]

chromosome. The genes at that one locus are called *alleles*. If one of the alleles has mutated, while its partner remains unchanged, the animal carrying one of each kind is said to be *heterozygous*. When a mutation has no visible effect in the *heterozygote*, it is called a *recessive* mutation. If it does show some modification of the heterozygote, it is said to be *dominant*. A recessive mutation can be carried along for several generations without any sign of it until, by chance, two heterozygotes, each carrying that particular mutation, are mated together. When that happens, about a quarter of their offspring will get from each parent the chromosome carrying the mutant gene. They are *homozygous* for it and will show its effects. In our example, the result was a puppy with bird-tongue.

Some of us find it simplest to think of the difference in appearance between the heterozygous state and the homozygous one as the outcome of a conflict between two opposing forces. If the recessive mutation is the weaker of those two, its action is suppressed in the heterozygote by the dominant allele. The heterozygote is thus protected by that dominant allele from any attempt of the recessive mutant to interfere with normal development, to change the color of the coat, to make the animal bigger or smaller, or to modify it in any way.

However, when there is no protective dominant allele for normal development because the homozygote has received the mutant gene from both sire and dam, the two recessive genes can combine forces to induce whatever changes they can cause. In some cases, the effect is no more serious than a change in color; in others, as in the bird-tongue puppies, it can cause death. Such *lethal genes* may exert their deadly effects very early in embryonic development (as in mice homozygous for yellow coat, and horses homozygous for dominant white), at birth (as in bird-tongue puppies and many lethal conditions in other domestic animals), at various ages after birth (as with some kinds of ataxia in dogs), or even not until middle age, as with Huntington's chorea in man.

There is one other adjective that can be applied to the gene causing bird-tongue. It is an *autosomal* gene; i.e., located in one of the 38 chromosomes other than the pair that determines sex. This was clearly shown by the fact that both males and females were affected and with approximately equal numbers in both sexes.

Thus, in full definition, the bird-tongue defect is caused by a simple recessive, autosomal gene that is lethal to homozygotes soon after birth.

Dominant Mutations

Dominant mutations are less common than recessive ones, but they do occur. In dogs, one of them is a kind of hairlessness that has been adopted as a breed character to make breeds known in different parts of the world as Chinese, Turkish, Persian, or Mexican Hairless.

Another is a defect designated by Patterson et al. (1967) as *congenital hereditary lymphoedema*. It is manifested as a swelling of soft tissues and caused by excessive accumulation of fluid, which, in turn, is attributed to defects in the lymph system. The hind legs are usually affected more than other parts of the body. Some of Patterson's dogs with extensive oedema died before weaning. In others, there was so little swelling that the condition was difficult to detect. Puppies mildly affected usually survived to maturity and appeared healthy, but breeding tests showed that they could transmit the defect to their progeny, just as dogs severely affected did.

As with some dominant mutations in other species, the *expressivity* of the *character* is evidently extremely variable in the heterozygote. In genetic parlance, the character is the effect induced by a gene. It is usually visible, but can sometimes be detected only by special tests. Expressivity refers to the range in visibility of the character.

Breeding Tests

Dominant mutations cannot be distinguished from recessive ones by their appearance, but only by breeding tests. To explain these tests, a few more terms may help.

Instead of referring to specific genes as "the gene causing bird-tongue in the dog" or as "the gene causing lymphoedema," geneticists use their own shorthand, a *symbol*, which saves much time and space. Thus, we might label the gene for bird-tongue *bt* and that for lymphoedema *Ly*, using capital letters only to designate dominant mutations, and trying to pick a symbol that suggests the full name of the mutant gene and of its normal allele. Recognizing that the genes occur in alleles, we can designate the *genotypes* (genes carried) and *phenotypes* (appearance) for the two mutations considered thus far, as in Table 1-1.

TABLE 1-1 Genotypes and Phenotypes

	Genotype	Phenotype
For bird-tongue (recessive)		
Homozygous, normal	*Bt Bt*	normal
Heterozygous	*Bt bt*	normal
Homozygous, mutant	*bt bt*	bird-tongue (lethal)
For lymphoedema (dominant)		
Homozygous, normal	*ly ly*	normal
Heterozygous	*Ly ly*	lymphoedema
Homozygous, mutant	*Ly Ly*	lymphoedema

With respect to bird-tongue, because homozygosity for the gene *bt* is lethal and because a mating of heterozygote × normal will yield only normal offspring, there is really only one kind of a mating that can tell us much. It is that of heterozygote × heterozygote *(Bt bt × Bt bt).* As the animals are phenotypically normal, we cannot know whether an animal is a heterozygote (commonly called a *carrier*) or not, until puppies with bird-tongue appear in some litter. Then we know that both sire and dam are *Bt bt.* The proportion of bird-tongue puppies in their progeny should be about 25 percent. For the present, the easiest way to recognize how that figure is determined is to remember that the chance of any one pup getting *bt* from his sire is 1 in 2; from his dam, 1 in 2; and from both, 1 in 4.

In twelve such litters from parents both carriers there were 76 pups altogether, so, if the defect were a simple recessive trait, the ratio theoretically expected was 57 normal : 19 bird-tongue. Actually, the ratio observed was 54 : 22. The slight excess of affected puppies is not significant, but, as we shall see later on, such an excess is to be expected when ratios are counted for only the litters (or families) in which at least one pup is affected.

We should remember that when the first puppy with bird-tongue appeared there was no previous record of such a condition, hence no way of knowing whether it was inherited or not. The second one in the same kennel shed a little light on that situation, but it was not until enough of them had been born to demonstrate a clear 3 : 1 Mendelian ratio that the defect could be attributed to action of a single recessive gene in the homozygous state. If the mutation recurs in some other breed,[1] its genetic basis will be known at once, and its elimination should be simple.

As for lymphoedema, as soon as the first affected dog produced similarly affected offspring, it was recognizable as hereditary and dominant. That a single dominant gene was responsible was attested by five matings of affected × unaffected *(Ly ly × ly ly),* which yielded 40 puppies. Among these, Patterson et al. (1967) found 20 with lymphoedema and 20 without. Such a perfect fit of observed numbers to the 1 : 1 ratio expected is not likely to occur very often in such large numbers, but it can happen.

In short, dominant mutations are passed from one generation to the next and are clearly visible in both. A recessive gene can be passed along unknown for several generations until by chance two carriers are mated, at which time about a quarter of their offspring (if there are enough of them) should show the trait that it induces.

---

[1]Yes, I know you'd like to know the breed, but I promised the owner not to tell.

Incomplete Dominance

A dominant mutation causing a clearly visible effect in the heterozygote may induce still greater changes in some of the progeny when two such heterozyotes are mated together. By chance, about 25 percent of the offspring will get that dominant gene from each parent. They will thus have a double dose of a gene with powerful effects. In some cases, the modifications induced are so extreme that the animal is unable to survive and dies during gestation or soon after birth.

A good example in dogs is the extreme hairlessness (hypotrichosis) briefly referred to earlier. According to Letard (1930), when two hairless dogs are mated together, their homozygous progeny have such extreme abnormalities (absence of external ears and malformation of the buccal cavity) that few are born alive and none survive. Accordingly, every dog with that dominant type of hairlessness must be heterozygous. It seems probable, however, that there are other kinds of hereditary hypotrichosis which are not lethal to homozygotes.

A more familiar example of incomplete dominance is provided by the merle type of dilution of pigment found in merle Collies, calico Foxhounds, the Norwegian Dunkerhund, Harlequin Great Danes, and dappled Dachshunds. The fact that this kind of dilution has been adopted as the distinguishing feature of several breeds and varieties shows that the gene responsible for it makes a very attractive dog (Figure 1-2), but it does so only for the heterozygote. Merle dogs have

FIGURE 1-2
A heterozygous blue merle Collie showing irregular blotches of black on a background that is bluish gray and diluted tan. [From Mitchell, 1935.]

black (undiluted) blotches on a background that is either dilute tan or bluish gray. In homozygotes, the double dose of the merle gene causes serious defects. Such animals are usually almost entirely white, with small, pale blue eyes, often blind, and occasionally lacking one eye or both (Figure 1-3). Moreover, as in other blue-eyed white mammals, most of them are deaf.

The attractive merle pattern and color is caused by an incompletely dominant gene, *M*, in the heterozygous state (Mitchell, 1935; Sorsby and Davey, 1954). Matings of *Mm* × *Mm* yield three distinct types—not merle, merle, and extreme (defective) merle in the ratio of about 1 : 2 : 1, respectively (Table 1-2).

This is not the only case in which breeders of domestic animals have liked the phenotype of the heterozygote so well that they have set up as their standard, ideal type a genotype that cannot breed true. When mated together (i.e., within the breed) such heterozygotes yield progeny among which only about half conform to breed standards and the other half do not. Familiar examples are Palomino horses and Blue Andalusian fowls. There are even breeds in which the homozygous condition is lethal to the 25 percent that get a double dose of the gene distinguishing the breed; e.g., Creeper fowls, Platinum foxes, Blufrost minks, Grey Karacul sheep, and Dexter cattle.

One must not get from these examples any notion that a gene incompletely dominant in the heterozygote is always disastrous to the homozygote that has two such genes. Sometimes animal breeders are so broad-minded that they accept without demur all three phenotypes

FIGURE 1-3
A homozygous merle Collie, mostly white, bred in France, owned by K. and W. Reinhard, Oderheim, Germany. [Photo by Eva-Marie Vogeler in *Collie Revue*.]

TABLE 1-2 Inheritance of the Merle Pattern

Parents and Breed	Progeny		
	Not Merle	Merle	Extreme Merle
Merle × not merle			
Dachshund	102	97	0
Collie	24	21	0
Miniature Collie[a]	16	12	0
	142	130	0
Merle × merle			
Dachshund	5	11	10
Collie	2	5	3
Miniature Collie[a]	7	7	0
	14	23	13

[a]Probably Shetland Sheepdog, but this is the name used by Sorsby and Davey.
Source: Sorsby and Davey, 1954.

that result when such heterozygotes are mated together. Roan Shorthorn cattle have a very attractive mixture of red and white hairs. Roan × roan yields approximately 1 red : 2 roan : 1 white, and all three colors can be registered as purebred Shorthorns.

Many dominant genes are eventually found on further study to be only incompletely dominant. Sometimes the manifestation of the dominant character diminishes so much with repeated outcrossing to unrelated stock that the proportion of animals actually showing it falls below the 50 percent expected in such matings. This happens because the unrelated stock is likely to contribute "modifying genes" that suppress (in varying degrees) the action of the dominant gene. Such modifiers are considered in Chapter 2.

Those Mendelian Ratios

Dog breeders do not really have to bother much about Mendelian ratios, but they should know why those ratios are important to the geneticists who seek to find what characters are inherited, and how. To illustrate the standard procedures they would use, let us suppose that a breeder of black Cocker Spaniels[2] has asked some geneticist to explain the unexpected appearance of a brown pup in a litter from parents both black. The answer has been known for years, but the case will serve nicely to show what kinds of matings are desirable to find

[2]One reviewer has taken me to task (here, and in 20 other spots) for not specifying whether the Cockers of which I write are American Cocker Spaniels or European Cocker Spaniels. Since inheritance of color is the same in both, any such distinction matters not. Readers can assume, however, that the Cockers in Sweden and Switzerland (Chapter 12) are more likely to be European than American!

out how the brown color is inherited. They, in turn, will introduce us to some more terms in the geneticist's vocabulary.

At maturity of the brown dog, it would be crossed to black ones, preferably unrelated. When none of the progeny are brown, we know that brown is recessive to black, if it is inherited at all. Expectations for a recessive trait in subsequent matings are illustrated by the diagrams shown in Figure 1-4. The terms P_1 and F_1 used therein to designate successive generations should be noted (P stands for *parental,* F stands for *filial*).

Assuming that the difference between black and brown depends on a pair of allelic genes, *B* (black) and *b* (brown), and that the P_1 black is *BB* and the brown *bb,* every F_1 dog must get one of the pair from each parent and must, therefore, be *Bb.* Known carriers such as these are commonly designated in pedigree charts by a dot in their pertinent squares or circles.

The next step is to mate some of these black F_1 heterozygotes with the unusual type, that is, brown. With the *B* and *b* alleles, it is not essential that the original brown dog be used for this mating. Other brown Cocker Spaniels will do equally well, as they, too, should have the genotype *bb.* This type of mating ($F_1 \times$ recessive) is called a *backcross,* or *test-cross.* Because all of the backcross progeny will get *b* from their brown parent, but only about half will get *B* from their black *(Bb)* parent, the ratio expected is 1 black : 1 brown.

While such a backcross usually tells the story, to make sure that only a single recessive gene is involved, some of the F_1 animals should also be mated *inter se* (among themselves) to produce an F_2 generation. It should show (approximately) the classical 3 : 1 Mendelian ratio now known to almost every highschool student who has studied introductory biology.

In the backcross, all of the black dogs must be heterozygous, but in an F_2 generation they are not. When *gametes* (reproductive cells, either eggs or sperm) are formed in the F_1 hybrids *(Bb),* the paired chromosomes separate, and one of each pair goes to each gamete. Accordingly, half of every female's eggs carry *B,* and half *b.* It is the same with the sperm cells. Assuming random fertilization of female gametes by male ones, the resulting genotypes in the F_2 generation are about 1 *BB* : 2 *Bb* : 1 *bb* (Figure 1-5). The last of these three is recognizable by its brown color, but the first two are indistinguishable except by a further breeding test; i.e., black is completely dominant to brown. Another way of saying the same thing is that, while the phenotypic ratio is about 3 black : 1 brown, approximately only one of the three F_2 blacks will breed true to type, and two could produce browns when mated to other *Bb* black dogs or to brown ones.

Finally, there is still one more kind of mating that can be used to clinch our conclusion that the brown color is caused by a simple reces-

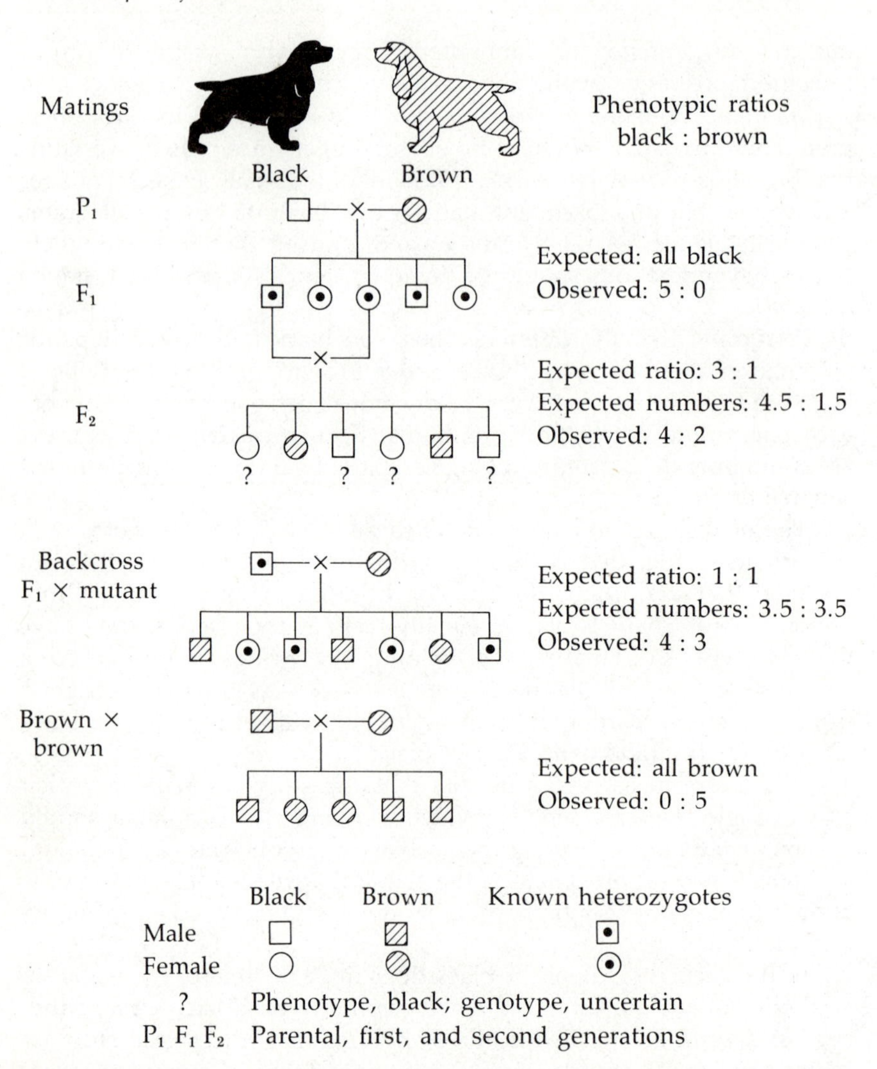

FIGURE 1-4
Inheritance of black and brown in various matings of Cocker Spaniels. The blacks are shown with clear symbols so that those carrying the gene for brown can be identified.

sive gene. It is the mating of brown × brown. Because neither parent carries *B*, all resulting progeny should be brown.

Interaction of the *B* and *b* alleles with other genes (*E* and *e*) causes some *bb* dogs to be brown, liver, red, or even yellow, as we shall see in the next chapter.

"About and "Approximately"

The words *about* and *approximately* have been used several times in the foregoing paragraphs to warn the reader that the exact ratios expected

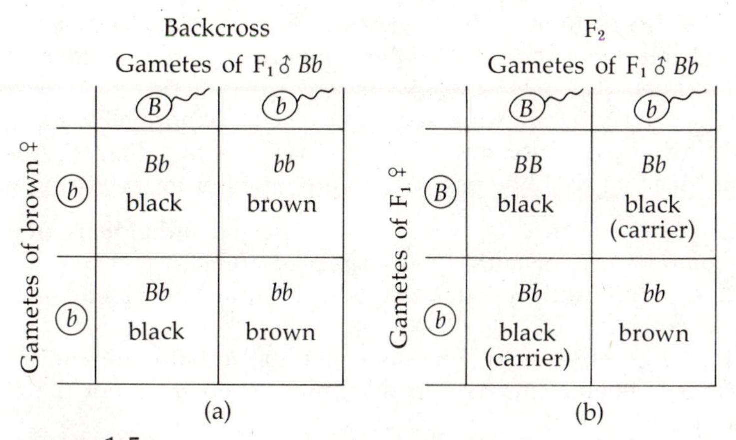

FIGURE 1-5
Random fertilization in (a) a backcross of an F_1 ♂ (*Bb*) to a brown ♀ (*bb*), and (b) a mating of F_1 ♂ × F_1 ♀ to beget an F_2 generation. The former yields a ratio of 1 : 1; the latter, 3 : 1.

are seldom actually observed, especially in small numbers. For expected ratios, we assume random fertilization. This means that (for example), in a backcross of *Bb* × *bb*, just as many gametes carrying *b* should fertilize eggs as do gametes carrying *B*. However, those gametes are never counted out in equal numbers, and, by chance alone, there can be deviations from the 1 : 1 ratio expected. There are ways to determine whether or not such discrepancies are statistically significant or not. If they are, then further study is necessary to find out why one is deficient. It is not uncommon to discover that deficiency of the recessive type results from death of some of the homozygotes before birth or soon after birth before they are counted. Tests for the significance of discrepancies between observed and expected ratios are discussed in Chapter 6.

To emphasize the fact that deviations of observed from expected numbers are bound to occur in small numbers, the observed F_2 ratio in Figure 1-4 is shown as 4 : 2, compared with the 4.5 : 1.5 expected for 6 F_2 animals. That is a good fit, but by chance alone there might have been 5 : 1 or even 3 : 3. If we had 100 F_2 animals (a common occurrence with mice or chickens), the observed numbers would fit more closely the expected 75 : 25. Similarly, the observed ratio of 4 : 3 in the backcross is as close as we could come (in whole dogs) to the 3.5 : 3.5 expected. By chance alone, there could have been 5 : 2, or even 6 : 1.

Genetics in the Kennel

Experimental matings of the kinds described in the previous paragraphs are fairly easy to make in mammals no bigger than a rabbit, but they are hardly practicable in hounds or horses. Nor are they necessary. We do not go in for backcrosses and F_2 ratios in *Homo sapiens,* but we probably now know more about inheritance in man than in any other animal.

When a puppy showing some unexpected abnormality appears in the kennel, the question immediately arises "Is it hereditary?" Answers to this question could be provided in any or all of the following ways:

1. The occurrence of two or more affected individuals from the same dam or sire, or both, or from related animals.
2. A significantly greater frequency within one breed, or strain, than in others.
3. Pedigrees showing ancestors common to both sire and dam.
4. Experimental matings to determine whether or not the abnormality recurs.

Some of these indicators are more conclusive than others. A pedigree showing inbreeding tells only that the abnormality might be inherited, but does not prove it. Experimental matings are seldom necessary.

As a first step, the breeder should find out whether or not the abnormality is already known to be hereditary. Fortunately, knowledge about heredity in dogs is steadily increasing, and the list of known genetic defects is lengthening. This book cannot give all the answers, nor can any of those sources listed in the references section at the end of this book. By the time any book appears in print, new defects and new facts about them will have been discovered since the last page of the manuscript was typed. A lot of "hush-hush" information circulates in breed societies before the facts get into print, so any dog breeder worried about some off-type abnormality in his kennel should consult not only the books and his veterinarian but also his friends, remembering always that there are some things about which "even your best friends won't tell you."

If the condition has not previously been reported in the literature on canine defects, or if genetic analyses have not been adequate to determine the mode of inheritance, experimental tests are warranted if the breeder has the facilities, the time, and the interest to make them.[3] In fact, one of the classical tests outlined earlier has already been made. When normal parents produce an off-type pup, one's first theory is that it may be showing a simple recessive mutation. Both parents are then suspected to be carriers, and their litter is equivalent to an F_2 generation. When other such litters are produced, either from the same parents or relatives, they should show about 3 normal : 1 abnormal.

Backcrosses tell us more than do F_2 generations, because in the former the recessive type should show up about twice in every four

---

[3]I have found that a little nudging by an interested geneticist helps, too.

offspring, whereas in an F_2 only about one in four is likely to show it. The dog breeder might be willing to raise one or more of the mutant type for use in backcrosses, because these are also test-crosses; that is, they can be used to test suspect dogs to see whether or not those suspects are carriers.

Sometimes (as with the bird-tongue defect) the mutant type cannot survive to reproduce. The only test-cross the breeder can make is that of suspect × known carrier. If the suspect is a carrier, the ratio expected is 3 : 1, and even a single mutant pup is enough to show that both parents are heterozygous for the mutation. However, if a litter of four does not show the abnormality, we must not assume that the 4 : 0 ratio proves the suspect to be innocent: It doesn't. If there are 12 pups in two such test-litters, and none shows the defect, we can be fairly certain that the suspect is not a carrier. The chance that he or she has fooled us is actually 0.0317, or 1 in 31.5, which is still a chance, but a small one. (For other probabilities, see Chapter 6.)

Complications

It should be recognized that not all *congenital* defects are genetic defects. The former term means only present at birth. It is generally understood to include not only any defect clearly visible at birth but also those caused by inheritance that do not show up until later in life. Some congenital abnormalities result—not from genes—but from accidents during development. Such defects are not passed on to later generations."[4]

Sometimes animals that should show some mutation fail to do so but are found by breeding tests to transmit it just as readily as do the animals in which the mutation is clearly evident. In such cases, the genotype is sometimes said to show incomplete *penetrance.* The degree of penetrance (measurable as a percentage) can vary with age, environmental influences, breed, and strains within a breed. Some geneticists do not like the term because it can be used to explain away differences between observed and expected numbers when some hereditary variation refuses to fit neatly into the Procrustean dictates of simple Mendelian ratios.

Some genes have *pleiotropic* effects; i.e., modifications of form or function other than the one first recognized. In other words, a single gene can induce a whole syndrome, the extent of which may be entirely unknown when the condition is first recognized as hereditary. For example, the assorted abnormalities in homozygous merle dogs are pleiotropic effects of the gene inducing that color pattern in heterozy-

---

[4]A fact impressed on one geneticist by 229 long-tailed rats, all of which failed to achieve the distinction of their congenitally tailless progenitor.

gotes. Similarly, when it was found that some gray Collies died prematurely, that lethal effect was first described as a pleiotropic effect of the gene for dilution of black to gray. Later it was found that the essential feature of the whole syndrome was a cyclic deficiency of certain cells in the blood. The phenotype is no longer simply gray Collie; it is a syndrome that includes cyclic neutropenia (see Chapter 10).

Other complications result from the interaction of one pair of alleles with another, as we shall see in Chapter 2.

2 Interactions and Exceptions

As the principle of Mendelian inheritance was gradually being found applicable in species other than garden peas, exceptions to the neat rule of 3 : 1 began to appear. We have already considered one of them—incomplete dominance. From study of other exceptions, it soon became evident that comparatively few genes are absolute autocrats among the thousands of others that determine inheritance in any one individual or species. The ultimate expression of any one gene depends in large part on its interactions with all the rest. Let us consider a few kinds of such interactions. For others in domestic animals, see Hutt (1964).

Complementary Genes

The *B-b* alleles of Cocker Spaniels behave as was shown in the preceding chapter only in the presence of another gene, *E*, which can be thought of as a dominant gene necessary for extension of the black or brown pigment throughout the coat. In *BB* and *Bb* dogs homozygous for *E*'s recessive allele *e*, black pigment is restricted to the eyes and nose. The coat is red. *B* and *E* are thus *complementary* genes, and both are necessary to make a dog black. The colors of dogs resulting from the interaction of the *B* and *E* alleles are as follows:

Genotype	Phenotype
BB EE *Bb EE* *BB Ee* *Bb Ee*	Black
BB ee *Bb ee*	Red, with black nose and eyes
bb EE *bb Ee*	Brown, liver, or chocolate, with brownish nose
bb ee	Yellow, pale red, or cream, with pinkish nose

These interactions of the genes *B-b* and *E-e* have recently been verified by Templeton et al. (1977), with color photographs to illustrate the resultant colors, in Labrador Retrievers. They classified reds (along with yellows) as yellow.

Dihybrids

The *E-e* alleles are independent of *B* and *b,* but both pairs of genes conform to the laws of Mendelian inheritance. In genetic parlance, each *segregates* in a clear 3 : 1 ratio. To see what happens when both are involved in the same animals, let us consider a cross between pure red *(BB ee)* and pure brown *(bb EE)* dogs.

The F_1 dogs will be black, but will be heterozygous in each pair of alleles, *(Bb Ee)* and are therefore called *dihybrids.* When they form eggs or sperm, it is entirely a matter of chance whether any gamete carries *BE, bE, Be,* or *be,* because the *B* alleles are independent of the *E* alleles. Therefore, when two such dihybrids are mated together, all four kinds of sperm cells have equal chances of fertilizing all four kinds of eggs. The resulting combinations in the *zygotes* (fertilized eggs) yield black, brown, red, and yellow dogs. The proportions in which they are likely to occur are most easily recognized by showing all possible combinations in a checkerboard, or Punnett square, so named after Professor R. C. Punnett, who originated it (Figure 2-1). It shows that there should be about 9 black : 3 brown : 3 red : 1 yellow.

That 9 : 3 : 3 : 1 ratio does not apply only to the coat colors shown in Figure 2-1. It is the standard ratio in any F_2 generation when two independent pairs of alleles are involved. It is not restricted to colors—or to dogs. For example, in bantam chickens the cross of rose-comb, black × single-comb, white produces only rose-comb blacks, but in the F_2 generation the ratio is about 9 rose, black : 3 rose, white : 3 single, black : 1 single, white.

Blacks and Reds

Dog breeders are not likely to see any such dihybrid ratios, but knowledge of the genes that cause them will make it easier to understand

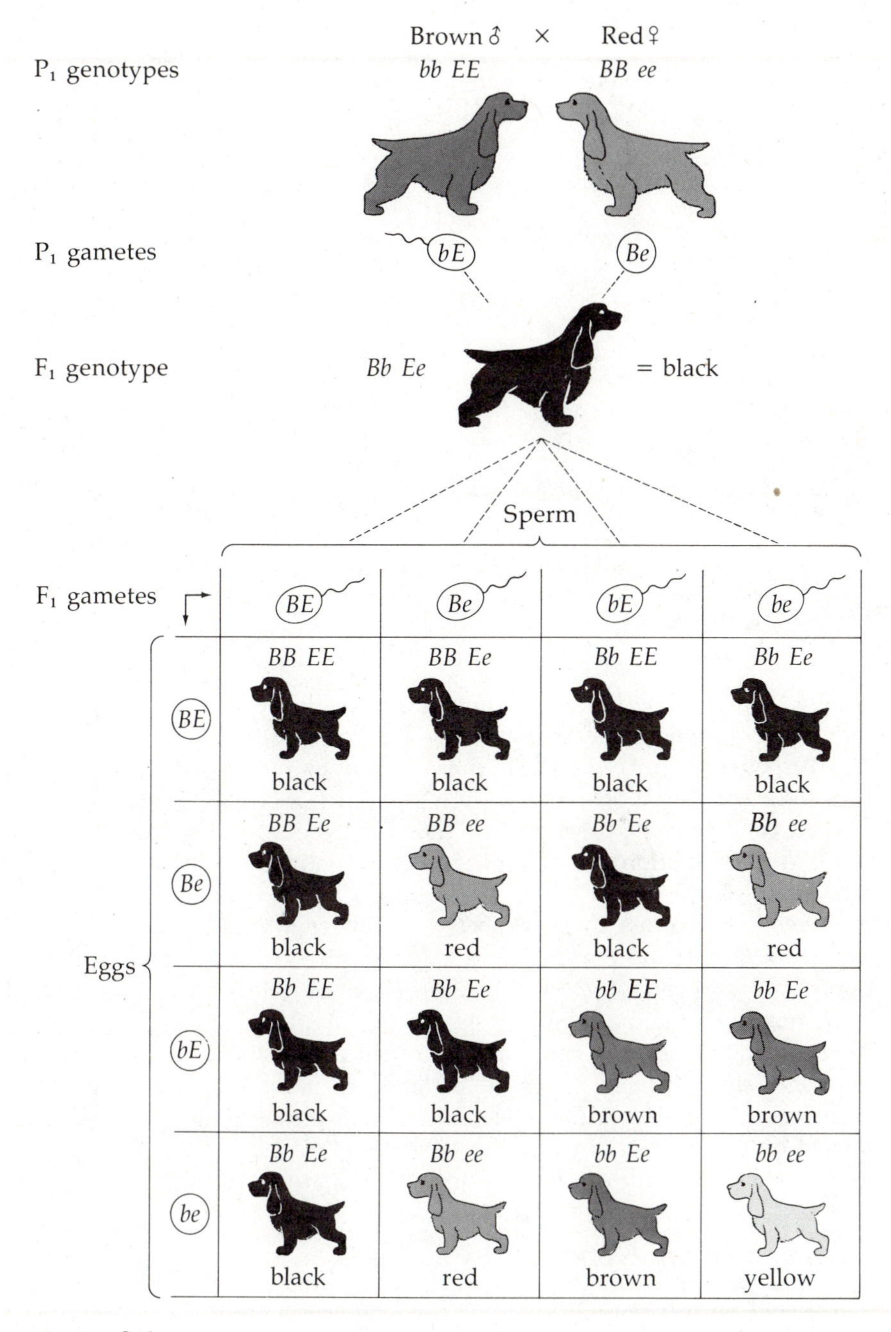

FIGURE 2-1
Independent inheritance of the *B-b* and *E-e* alleles, with interactions. The phenotypic ratio in the F_2 generation is 9 black : 3 brown : 3 red : 1 yellow.

TABLE 2-1 Colors Expected in Puppies from Matings of Black and Red Dogs of Differing Genotypes

Phenotypes and Genotypes of Parents	Phenotypic Ratio in Progeny
I. Black × black	
1. *BB EE* × any genotype	all black
2. *BB Ee* × *BB Ee* or *Bb Ee*	3 black : 1 red
3. *BB Ee* × *Bb EE*	all black
4. *Bb EE* × *Bb EE* or *Bb Ee*	3 black : 1 brown
5. *Bb Ee* × *Bb Ee*	9 black : 3 red : 3 brown : 1 yellow
II. Black × red	
1. *BB EE* × *BB ee* or *Bb ee*	all black
2. *BB Ee* × *BB ee* or *Bb ee*	1 black : 1 red
3. *Bb EE* × *BB ee*	all black
4. *Bb EE* × *Bb ee*	3 black : 1 brown
5. *Bb Ee* × *BB ee*	1 black : 1 red
6. *Bb Ee* × *Bb ee*	3 black : 3 red : 1 brown : 1 yellow
III. Red × red	
1. *BB ee* × *BB ee* or *Bb ee*	all red
2. *Bb ee* × *Bb ee*	3 red : 1 yellow

Source: Animal Genetics, by F. B. Hutt. Copyright © 1964 by the Ronald Press Co. Reprinted by permission of John Wiley & Sons, Inc.

some of the unexpected exceptions occasionally seen. That is particularly the case when interacting genes are involved.

It is obvious (in Figure 2-1) that black dogs may have any one of four different genotypes. Accordingly, matings of black × black do not always yield only black pups. All reds must be *ee,* but some are *BB,* other *Bb.* Expectations in different kinds of matings involving black or red are shown in Table 2-1.

Obviously, colors to be expected in different matings depend on the genotypes of the parents—not on their phenotypes. And, as Mating I-3 shows, even when black parents produce only black pups, there is no guarantee that both parents are *BB EE.* Similarly, Matings B-2 and B-5 both yield black and red pups in the ratio of 1 : 1, but that does not tell completely the genotypes of the parents. Nevertheless, when we know how the *B-b* and *E-e* alleles interact, parental genotypes are revealed by the progeny in 9 of the 13 matings of Table 2-1. Those nine provide in simple form a *progeny-test* of the parents.

Complementary Genes and Paralysis

A peculiar kind of paralysis of the hind limbs was found by Stockard (1936) in offspring of the cross Great Dane × St. Bernard. It began at about three months of age, with symptoms varying from slight paralysis to inability to walk unaided. This is quite different from the hip dysplasia so prevalent in big dogs. The trouble is in the nervous system, specifically a loss of motor neurons in the lumbar region of the spinal cord. That causes paralysis of the muscles of the thigh, except the *sartorius,* which is the chief flexor. Attempts to compensate for their paralysis result in an ambling gait and rotation of the hind legs, but the abnormality is not lethal.

It occurred only occasionally in pure Great Danes and St. Bernards, but, in 57 F_1 progeny from the cross between these breeds, all developed paralysis except 3 or 4. Among 66 dogs in an F_2 generation, 20 became paralyzed. From these facts, it was concluded that the paralysis resulted from the interaction of complementary genes from both parent breeds. Few dog breeders are likely to cross these two breeds, but all are advised not to breed any dogs of either breed that show this kind of paralysis.

Epistasis in Irish Setters

Sometimes the expression of a dominant gene is prevented by *epistasis* ("standing on"); that is, suppression of one gene by another at a different locus, perhaps even in a different chromosome. This is different from simple dominance of one allele over its partner at the same locus. Winge (1950) has postulated that such an epistatic gene *(A)* is responsible for the rich, mahogany red of the Irish Setter in many dogs of that breed. They carry black. That black is not recessive to a dominant red; it is *hypostatic* to epistatic red. Winge states that these concealed blacks can be recognized by black in the tip of the nose and on the balls of the feet. In Irish Setters not carrying black, these areas are brown.

A familiar example of epistasis is provided by the dominant white plumage of White Leghorns. Those chickens would be black and barred, were it not for a gene that inhibits formation of black color, and hence of any pattern in that color. The gene for dominant white *(I)* is dominant to its recessive allele *(i),* but epistatic to *C,* a gene for black pigment in the feathers.

Linkage

Sometimes two genetic traits that one might suppose to be quite independent of each other stick together in crosses and do not segregate in Mendelian ratios. That happens when the genes involved are on the same chromosome. Fortunately dog breeders need not bother to learn the esoteric details of *linkage,* because, at the time of this writing, only four genes are known to be linked in dogs. We shall come to them in Chapter 4.

Nevertheless, we should know what linkage is all about, because it can be important if a gene for some undesirable defect should prove to be linked with one causing a trait that is wanted. In all species, the numbers of genes far exceed the numbers of chromosomes. Clearly, most chromosomes must carry many genes. There are techniques by which such linked genes can be recognized and by which one can determine whether they lie closely together or far apart. Some such *linkage groups* can eventually be assigned to identifiable chromosomes, and for some species there are *chromosome maps* showing relative locations of known linked genes. Such maps can be helpful in breeding for desired objectives, but, for the dog, any such map lies a long way in the future.

This is no reflection on dogs, or their owners. The map best known is that for the fruit fly, Drosophila, which can produce a new generation in about two weeks. Among mammals, fame for the most extensive map belongs to the mouse. Obviously the mapping of chromosomes proceeds fastest in animals that multiply rapidly.[1] Breeders can help to accumulate information about the dog's genes by reporting to interested geneticists any cases in which two different characters appear to be associated.

Multiple Alleles

Continuing with exceptions to the simple Mendelian rule that genes come and go in pairs, of which only one partner can occupy a specific locus on a chromosome, we come now to *multiple alleles,* a whole series of genes that can occupy a single locus. This is not any drastic reversal of ideas earlier given as facts—it is merely an extension of them. Chemically, a gene is a very complex structure. If a mutation can occur at one point in that structure, it can also occur at another. Two such mutations would result in three alleles at one locus. In some loci, particularly in those determining blood antigens in mammals, many different alleles have been identified, and at the *B* locus in cattle the number is well over 300.

Some tenets of our earlier faith remain! In any one animal, there can be only two alleles at one locus, no matter how many others are known in the series to which those two belong. It is the species that has multiple alleles—not the individual. An individual can have any two of the series and can be homozygous or heterozygous at the locus concerned. Multiple alleles affect only one character, causing different degrees of its expression. Usually they can be arranged in order of dominance, with every one dominant to those below it, and recessive to those above. However, in some series of multiple alleles, any two members have the equal rights now so fashionable, and in heterozygotes those two are both expressed. This is usually the case at loci that determine blood antigens. Such alleles are said to be *codominant*.

---

[1]And not a single Mendelian character has yet been reported in elephants.

Multiple alleles are designated by a superscript in their symbol. According to Little (1957), there are at least four series of multiple alleles affecting coat color in dogs. One of these (for example) has effects as follows:

Gene	Effect	Example
A^s	Extends black pigment over the entire body	Black Cocker Spaniel Black Labrador Retriever
a^y	Replaces black with sable or tan	Basenji, Collie
a^t	Bicolor. Black with tan on feet, muzzle, and above the eyes	Fox Terrier, Beagle, Doberman Pinscher

There may be other alleles in this series. For most dog breeders, multiple alleles will cause few worries or none, but the knowledge that such series can occur may be helpful in some cases, particularly when breed standards for colors and patterns are concerned.

Lethal Genes and 2 : 1 Ratios

Sometimes an expected Mendelian ratio of 3 : 1 is converted to 2 : 1. This happens when homozygosity for some gene is lethal. If the lethal action occurs early in gestation, the homozygote is resorbed and is never seen. The first lethal gene to be identified in animals was that causing yellow coat in the mouse. Matings of yellow × yellow always yield smaller litters and ratios of 2 yellow : 1 not yellow. Eventually it was found that the homozygous yellows all die at an early stage of gestation. Apparently the same thing happens with dominant white in horses.

If the lethal effect is fatal at later stages, the homozygote may be aborted prematurely, or die soon after birth (as with the bird-tongue lethal in the dog), or even survive for several weeks or months. In these cases, the abnormalities induced by the double dose of the gene can be recognized, so that phenotypic ratios observed will agree with those expected. As an example in the dog, from the mating of (dominant) hairless × hairless the expectations are 1 aborted, abnormal : 2 hairless : 1 normal. From parents both heterozygous for the bird-tongue lethal, the expectation is 3 normal : 1 bird-tongue, and that ratio is recognizable within 48 hours of birth.

Many different lethal genes are known in domestic animals, but those identified thus far in the dog are comparatively few. They are listed in Appendix I and will be described in the chapters dealing with genetic defects.

Modifying Genes

Some simple dominant or recessive characters that should segregate in clear Mendelian ratios fail to do so because of the action of *modifying genes*. These are genes with small effects, and in numbers unknown, which act cumulatively to enhance or to suppress the trait that they modify. Like the *polygenes* (to come in Chapter 5), the modifiers give the breeder that variation necessary for any effective selection toward objectives desired.

A good example is seen in the white spotting of Beagles. According to Little (1957), the piebald pattern of that breed is induced by a gene (s^p) that is one of the alleles at the *S* locus. Acting along with that gene, there are "plus" modifiers, which increase the extent of pigmented areas, and "minus" modifiers, which decrease them. He made 10 arbitrary grades of the piebald pattern ranging from darkest to lightest (Figure 2-2). All of these must have carried the s^p gene, but the variation in the amount of white was attributed to the action of modifiers. By selection, the breeder could accumulate such modifying genes to make his Beagles darker or lighter if he wished to do so.

Apart from color, other mutations can be suppressed by modifying genes. This is best recognized when animals carrying some dominant gene are outcrossed to unrelated normal stocks. At first the mutation may segregate nicely in the 1 : 1 ratios expected, but, if such backcrosses are continued, the proportion showing it may sometimes decline so much that the once dominant mutation is almost completely suppressed. In wild animals, any undesirable dominant mutation that has an adverse effect on viability could persist only by multiplication of the individuals least affected (that is, most modified). Eventually, by the accumulation of modifying genes, mutations originally dominant could become recessive.

To the breeder of domestic animals, all of this means merely that genetic mutations do not always have to show up in nice, sharply-defined classes and ratios. Sometimes animals homozygous for a recessive mutation may intergrade with normal individuals, so that the two genotypes can be identified only by a breeding test.

Different Genes with Similar Effects

In many species, the same phenotype is induced by different mutant genes. For example, as this is being written, some of my veterinary colleagues are trying to determine whether or not the hereditary ataxia in one breed of dogs is genetically the same as the hereditary ataxia in another. The age at onset and effects are much alike. The standard test to answer the question is to mate the two kinds together, but in this case that is not possible, partly because the two breeds are separated by oceans, and still more by the fact that few of the ataxic pups survive to breeding age.

When independent recessive mutations appear to induce identical phenotypes and when the two kinds are crossed, all the progeny

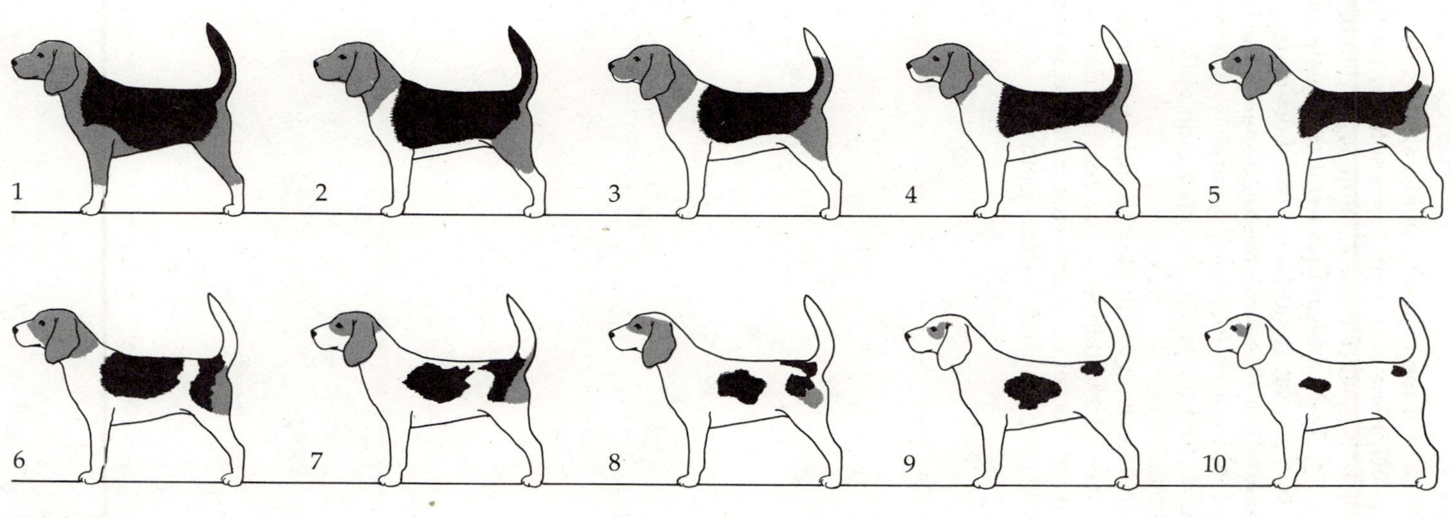

FIGURE 2-2

Ten grades of piebald spotting in Beagles. All the dogs carry the causative gene s^p, but variations in the extent of the colored areas are attributed to plus and minus modifying genes. See text. [Reprinted from Clarence C. Little, *The Inheritance of Coat Color in Dogs*. Copyright © 1957 by Cornell University. Used by permission of Cornell University Press.]

should show the mutation if the causative genes are identical. If those genes are at different loci, none of the progeny should show the mutation in the F_1 generation, but, since those F_1 animals are dihybrids, they should yield a ratio of 9 : 3 : 3 : 1 in an F_2 generation. Castle found this to happen with his two unrelated stocks of rex (short-haired) rabbits, but, because the two kinds of rex were indistinguishable, the last three classes in that ratio could not be identified separately. The 55 normal and 33 rex that he got fitted nicely the expected ratio of 9 : 7 in his F_2 generation of 88 rabbits.

Independent mutations with identical or similar effects can occur in any species. There may be two kinds of red coat in Irish setters (as Winge postulated) and in Cocker Spaniels. That could explain the occasional occurrence of black pups from parents both red.

3 Chromosomes and Genes

We all know now that what we inherit is passed from one generation to the next by genes, that these are carried in the chromosomes, and that they, in turn, are passed along from parents to progeny in the eggs and sperm. There are thick books on this subject, but, fortunately, dog breeders don't have to read them unless they want to do so. There is little or nothing that the breeder can do about chromosomes except to try (by selection of breeding stock) to pack them with desirable genes, and to eliminate from the kennel (again by selection) those that carry deleterious genes.

Nevertheless, we should all know a little about chromosomes and genes, if for no other reason than the fact that they do the work.

Chromosomes

The word *chromosome* means "colored body" and was assigned because that is how chromosomes appear under the microscope in cells specially treated and stained. In most cells of the body, the dog has 78 chromosomes (or 39 pairs), but in the germ cells (eggs and sperm) the number is reduced to 39. Details of the processes by which paired chromosomes part to go their separate ways into the definitive germ cells can be found in almost every good text on biology and need not be repeated here. The total number of chromosomes is called the *diploid* number, commonly designated as $2n$; that in eggs or sperm is the *haploid* number (n). At fertilization, the diploid number is restored.

A Japanese cytologist, Minouchi (1928), was the first to establish 78 as the diploid number in the dog. That same number was also found by Ahmed (1941) in the Sealyham Terrier and Spaniel, by Moore and Lambert (1963) in the Beagle, and by Gustavsson (1964) in the Dalmatian, Rottweiler, Border Collie, Airedale, and German Shepherd. Since four investigators, in four widely separated countries, and using constantly improved techniques, have found the same number of chromosomes in every breed examined, it seems reasonably certain that the standard diploid number of chromosomes is 78 in all breeds of dogs.

To prepare those chromosomes for study is a special art. To get the remarkably clear views of them shown in Figure 3-1, Gustavsson grew blood cells in tissue culture for three days with phytohemagglutinin added to promote cell division, then arrested such division with colchicine, washed the cells with distilled water, fixed (killed) them with one chemical, and stained them with another. Finally, the chromosomes were spread out in a "squash preparation."

The upper half of Figure 3-1 shows the chromosomes as they are seen under the microscope. To match them up in pairs, a photograph is taken, and each chromosome is cut out by itself and paired up with another of the same size. In addition to differences in length, chromosomes differ in location of a special part called the *centromere,* which is the point at which a spindle-fibre is attached when the cell divides. However, the centromere is not any help in pairing up the dog's chromosomes because (with one exception) they all have that centromere at one position only—at the end of the chromosome or very close to it. The exception, shown in the larger of the pair in the lower right corner is the sex chromosome, commonly called the X *chromosome.* Females have two of these, but in males there is only one, and its partner, oddly enough, is the smallest of the whole 78. It is called the Y *chromosome.* In X chromosomes (of the dog), the centromere is near the middle, so that the chromosome seems to have two arms. The Y chromosome differs from the X chromosome not only in size but also because its centromere is at one end or near it, and, above all, by the fact that it is never found in females.

When all the chromosomes have been matched, the 39 pairs are arranged in descending order of size to make a *karyotype.* It is clear that it must be difficult in many cases to decide from visual inspection exactly where a chromosome fits into a long series in which so many pairs appear identical in length. Sometimes the task is facilitated by weighing the cut-out chromosomes on a very sensitive balance, so that they can be paired by weight. A later method (called *G-banding*) uses special techniques to reveal transverse bands that differ in intensity of staining, and which occur in different patterns in different chromosomes. As Selden et al. (1975) have shown, these help to match up the chromosomes in pairs (Figure 3-2).

Chromosomes other than the sex chromosomes are called *autosomes.* The genes they carry are referred to as *autosomal;* those in the big sex chromosome are *sex-linked.*

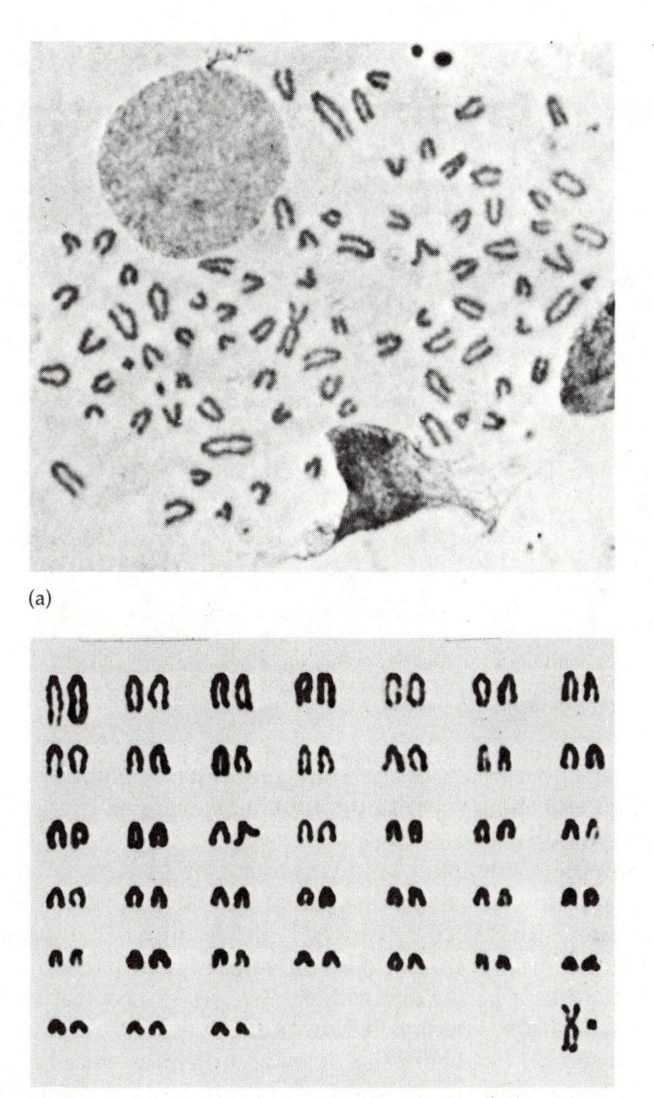

(a)

(b)

FIGURE 3-1
Mitotic chromosomes from a leukocyte of a male dog. (a) As seen under the microscope. (b) Arranged into a karyotype, with autosomes in order of diminishing size. Note X (larger) and Y (smaller) chromosomes at lower right. [From Gustavsson, 1964.]

Aberrant Chromosomes

We used to be taught that all individuals of one species have the same number of chromosomes, but that day has passed. Chromosomes, like dogs, can go astray. Cytologists are finding abnormalities of various kinds in the chromosomes of every species adequately investigated. Some of them are lethal, and it seems probable that such aberrations

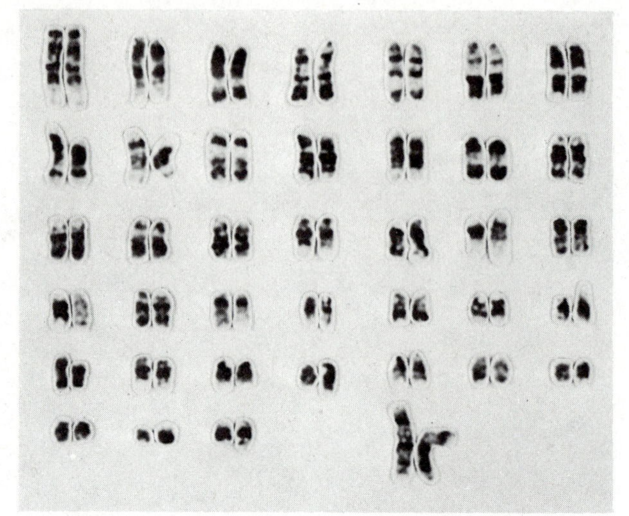

FIGURE 3-2
Karyotype of chromosomes from a female dog showing how matching of paired chromosomes is facilitated by the G-banding technique. Note that there are two X chromosomes (one bent) each about 5 microns long. A micron is one-thousandth of a millimeter. [From Selden et al., 1975.]

are responsible for much of the early mortality of embryos in various species. Some of them seem to do little harm or none to the individual carrying them.

An example of the latter kind was found by Ma and Gilmore (1971) in a crossbred dog that had no other detectable abnormalities. Counts of chromosomes in 300 cells showed a modal number of only 77, and further study revealed that one of the smallest chromosomes had become attached to a larger one (Figure 3-3). A similar fusion has been found in Miniature Poodles (Hare and Bovee, 1974). These abnormalities caused no trouble in the dogs, as no genes were lost or multiplied, but evidence in other species indicates that some *translocations* prevent formation of germ cells (when paired chromosomes should separate) and thus lessen the efficiency of reproduction.

Among 15 dogs with congenital heart defects, Shive et al. (1965) found one in which two chromosomes had fused to make one large chromosome with its centromere near the middle. Another dog had a minute extra chromosome. Because none of the other 13 dogs showed these abnormalities, they are evidently neither causes nor effects of congenital heart defects in dogs.

It is more serious when a whole chromosome is lost, or when three occur instead of the normal two. The best-known example of such an abnormality is probably Down's syndrome (mongolism) in man, in

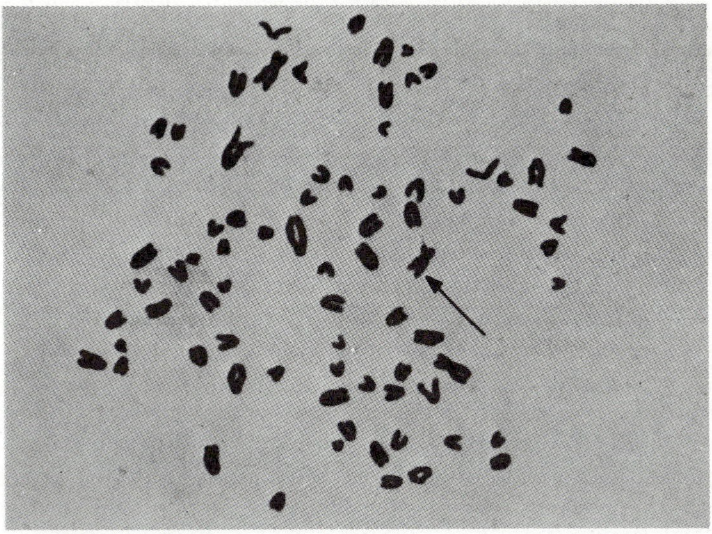

(a)

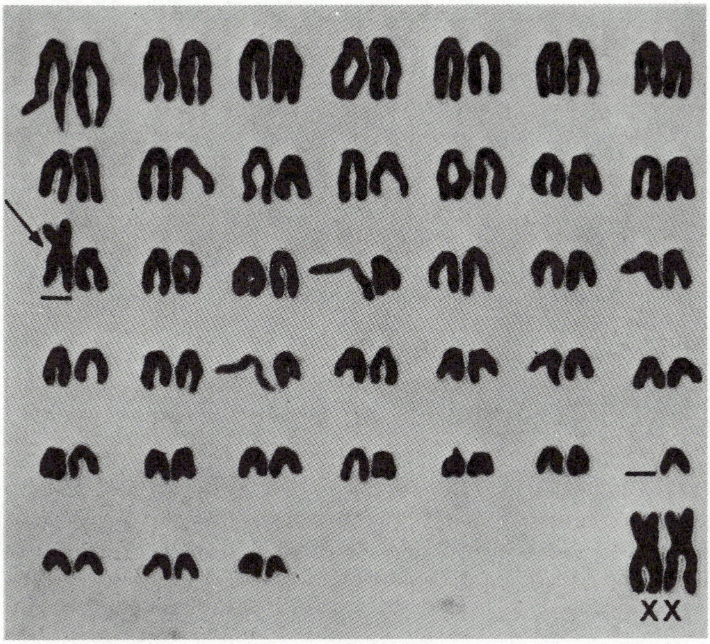

(b)

FIGURE 3-3

Chromosomes of a female dog, numbering only 77 because one of the smaller chromosomes had fused with a larger one. (a) As seen under the microscope. (b) Karyotype showing one unpaired small chromosome and the abnormal one resulting from the fusion. [From Ma and Gilmore, 1971.]

which there are three of one of the smallest chromosomes instead of two. A corresponding abnormality has not yet been found in the dog, but Clough et al. (1970) studied a male German Short-haired Pointer that had an extra sex chromosome. Its autosomes were normal, but the total count of chromosomes was 79, because each cell carried XXY instead of XY. Accidents of that kind result when paired chromosomes fail to disjoin during the formation of germ cells. This abnormality has now been found in several species of mammals, and in man it is known as the Klinefelter syndrome.

Clough's XXY male was sterile, as are other Klinefelters. The frequency of this condition in dogs is not yet known, but it should be suspected when a mature male with small testes is found to be sterile and no other cause can be found.

Thus far, few veterinarians have been trained to detect abnormalities of the chromosomes, but more will learn in the future. Any veterinarian with a small-animal practice should know the techniques necessary to obtain the cells for such an investigation, and, also, where to send them for study.[1]

It seems probable that more chromosomal aberrations will be recognized in the future in the dog, just as they have already been found in man and in other species.

Linkage

Since the dog has 39 pairs of chromosomes, there should be 39 *linkage groups;* i.e., groups of several (or many) genes all of which are in the same chromosome. To the best of my knowledge, only one such group is known in the dog, and it is in the sex chromosome, to be discussed in the next chapter. The autosomal groups will be harder to discover, because linkage is difficult to detect when the genes are scattered through 38 paired chromosomes.

Few dog breeders are likely to have problems caused by linked genes. They will have such problems whenever some lethal gene or other undesirable condition is proven to be linked with some character that should be retained. There is little point in discussing here the procedures for breaking up such linkages. Anyone wishing to pursue the subject further to learn how geneticists measure the closeness of linkage (i.e., how far apart the linked genes lie) or to see the relative positions of many genes in 19 linkage groups of the mouse and five groups in a map for the fowl can find a whole chapter on linkage elsewhere (Hutt, 1964).

In cases where two or more different inherited traits seem to go

---

[1]Having taught genetics to budding veterinarians for 20 years, I like to pass jobs to that profession whenever possible.

together, linkage of causative genes might be suspected. Special tests would be needed, however, to prove that the apparently linked characters are caused by separate genes, and are not part of a syndrome induced by one gene only. For example, dogs homozygous for the merle gene *(MM)* have predominantly white coats and defects of the eyes, are usually deaf, and often sterile. These assorted peculiarities are not caused by linkage of different genes. They are all pleiotropic effects of the gene *M*.

The Mighty Gene

None of us is likely to see a gene, but we do see the remarkable effects that genes can induce, so it won't hurt to know just a little about what a gene is like. Some calculations show that it could be from 0.0034 millimeters to twice that figure in length, but only 0.002 microns in width. (A micron is one-thousandth part of a millimeter.)

The gene is composed of deoxyribonucleic acid, commonly called DNA. The name of the acid comes from the deoxyribose sugar that it contains. Biologists now believe that a molecule of DNA consists of two long chains of nucleotides that coil slightly to form a double helix. Nucleotides, in turn, consist of a phosphate group, deoxyribose sugar, and either a purine or a pyrimidine.[2] Diagrams of the two chains of nucleotides (that double helix!) resemble a spiral fire escape, but some models of them look more like a long, narrow bunch of grapes, with constrictions at regular intervals.

Big genes can have lesser genes within them. These subgenes are known as *mutational sites*, and, according to one investigator, there may be several hundred of them at a single gene locus. Mutations at any one of these could affect in some way (perhaps a minor one) the genetic character influenced by the gene at that locus.

The ways in which these tiny genes induce the effects that we can see have become a special field of study for biochemists. Further details about them are unjustified in this book, but there are other books written by authors better informed than this one on that esoteric branch of science.

---

[2]If you must known about these, ask your biochemist to draw a picture of one for you.

4 Sex and Sex Linkage

About 1905, when it was established that sex is determined by certain chromosomes, some hundreds of theories about what makes boys and what makes girls were banished to limbo at one fell stroke. Some of the theories were ingenious, and some were titillating, but they had three things in common: (1) They all seemed plausible to their proponents, (2) they were all wrong, and (3) every one of them could be proven right half the time by chance alone (but so can any blind guess).

One of those theories that persisted over long among dog breeders was propounded (for man) about 450 BC by the Greek philosopher, Anaxagoras. It was very simple. Males are stronger than females, the right side of the body is stronger than the left; ergo, the right ovary must produce males. The left one, being on the weaker side, must produce females. The application (by dog breeders) was obvious. All one had to do to get more males was to restrain the bitch immediately after breeding and force her to lie on her right side for an hour or so until the spermatozoa had wriggled their way up into the right oviduct.

If this discriminatory theory had not been shot down (with all the others) by the chromosomes, it might have persisted up to the present were it not that in 1910 investigators found, after removal of either the right or left ovary from rats, that the remaining one yielded males and females in about equal numbers.

Determination of Sex

In the dog, as in other mammals, the male is *heterogametic;* that is, it produces gametes some of which are male-determining and some female-determining. Because the two kinds result when the X and Y chromosomes separate in an early stage of gametogenesis, there should be about equal numbers of both kinds. On the other hand, the female mammal is *homogametic.* Her eggs all carry an X chromosome, never a Y, and all are therefore equipotential so far as determination of sex is concerned.

Sex of any puppy is determined at fertilization and depends on whether that fertilization is done by a spermatozoon that carries an X, or by one that carries a Y. Because these two kinds are produced in equal numbers, the result should be a sex ratio of 1 ♂ : 1 ♀.[1] As we shall see later, accidents do occur, and the sex ratio in the dog is not exactly 1 : 1. In one way, that ratio is comparable to the 1 : 1 ratio expected when a heterozygote is backcrossed to a homozygous recessive. In each case, two kinds of gametes are produced in approximately equal numbers by one parent. The other parent contributes gametes all alike. After fertilization, there should be two classes of progeny in about equal numbers.

Although in all mammals the male is heterogametic, the female has that role in birds, moths, and butterflies. In some animals (reptiles, amphibians, and fish), males are heterogametic in some families, but homogametic in others.

Intersexes

Intersexes occur in dogs, as in other mammals, but are not common. In reviewing 48 such cases that had been reported in various breeds, Hare (1976) found that one-third of them were in Cocker Spaniels. The degrees of intersexuality varied from underdevelopment of external genitalia in males to hermaphroditism. A true hermaphrodite is an animal having both ovarian and testicular tissues, but in vertebrates one of those tissues is usually not functional.

In a unique case studied by Selden et al. (1978), an apparently normal Cocker Spaniel female was found to have bilateral ovotestes. These contained mature ovarian follicles and testicular tubules that were not producing spermatozoa. Her son was an intersex, with a uterus and with sex chromosomes XX instead of XY. From other evidence, including serological examination of the parents and a brother of the hermaphroditic female, it was considered most likely that her sire had acquired extra male-determining potency, either by mutation in an autosome (or the X) or by translocation of material from the Y chromosome. This potency had apparently been transmitted to her and by her to her son.

[1]To astronomers, these symbols—one representing Mars' shield and arrow, and the other the looking-glass of Venus—have for centuries meant Mars and Venus. We biologists borrowed them to designate a male and a female.

Sex Ratios

Having just shown that the sex ratio in dogs should be 1 : 1, we can now state that actually it is not exactly what it should be on theoretical grounds. As determined in several breeds by different investigators, *secondary sex ratios* (at whelping) vary somewhat, but nearly all show an excess of males. Some representative figures are given in Table 4-1. Others can be found elsewhere (Burns and Fraser, 1966).

There is little point in trying to establish any one figure for the species, or for any breed. As the data in Table 4-1 show, the ratios that have been determined by different investigators in the four breeds vary widely. It has not yet been proven conclusively that there are any significant differences in this respect among breeds.

Why so much variation? For one thing, the large numbers needed to offset chance fluctuation are not likely to be available in any one kennel, or to be recorded by any one observer. They must come from surveys (by questionnaire) or from stud books. In either case, there can be errors, and, certainly, the environmental conditions (including disease and management) would vary greatly. These last might affect the *primary sex ratio* (at conception) very little, or not at all, but they could influence the rates of subsequent prenatal and neonatal mortality. There is good evidence that such mortality falls more heavily on males than on females. Little's (1948) breeders of Cocker Spaniels found a sex ratio of 140.5♂♂ : 100♀♀ in pups born dead or nonviable. Wegner (1975) has reported consistently high proportions of males among stillbirths in German Shepherds. It is the same in most mammals (if not all), including man.

Whitney's (1939) data (Table 4-1) are exceptional, because they apply to his own kennel, to dogs kept under uniform conditions, and only to litters from dogs that had lived at least a year in that kennel. The sex ratio in his German Shepherds was 143 : 100 in litters conceived during the six colder months, but only 116 : 100 in those conceived from April to September. Such a seasonal influence has not yet been confirmed by others.

TABLE 4-1 Some Sex Ratios at Whelping, ♂♂ per 100♀♀

Breed	Number of Pups	Sex Ratio[a]	Authority
German Shepherd	1,440	124.3	Whitney (1939)
Cocker Spaniel	3,858	107.3	Little (1948)
Hungarian Sheepdog	5,848	105.0	Sierts Roth (from Burns and Fraser, 1966)
Bull Terrier	1,303	126.7	Briggs and Kaliss (1942)

[a]For others, see Burns and Fraser (1966).

Lest these somewhat deprecatory comments on surveys should discourage some dog owner from responding to others yet to be made, it is desirable to point out that surveys are very useful. Sometimes they point to real problems that need further study. Often they yield useful bits of information and suggest ideas for further research. For example, Little's (1948) breeders of Cocker Spaniels reported that black dams yielded a sex ratio (among 2,062 pups) of 113.5 : 100 but that among 1,796 pups from red mothers the sex ratio was only 99.5 : 100. Why such a difference? Statistical tests showed that it was "probably significant," but no reason why it should occur has yet been advanced.

It seems probable that in the dog, as in other mammals, including man, the primary sex ratio (at fertilization) is very high, even up to 150♂♂ : 100♀♀, but that subsequent mortality falls more heavily on the males, so that at birth they outnumber the females no more than is shown in Table 4-1. (In man, the corresponding figures commonly cited are 104–107♂♂ :100♀♀.)

Various reasons suggested for the high primary sex ratio in mammals include faster action by the Y-bearing sperm because of their presumed slightly smaller size, differences in survival of the two classes of sperm, and others, but none has yet been proven. Attempts to separate the two kinds by centrifugation and other methods have not been successful.

Similarly, no one reason can be assigned for the fact that prenatal mortality is consistently higher in males than in females. One theory about it is considered later in this chapter, and others can be found in other books. A former student of mine, who had an advantage over his teacher because he (the student) had taken courses in philosophy, solved it all very nicely with the declaration that it was clearly all arranged by Nature. It seemed only fair that the sex in greatest numbers at fertilization should be reduced more than the other before birth; otherwise (said he) the males would inherit the earth!

Enough of all this! More important to many a dog breeder is the question, "Why should that litter of five pups last week all be females when I have orders for three males? Could it happen again?" We answer that one in Chapter 6.

Sex-linked Genes

With rare exceptions in some species, the Y chromosome carries no genes, but the X chromosome always does. Obviously, any gene in the single X chromosome of the male will be transmitted only to the daughters; the Y chromosome goes only to the sons. As a result, sex-linked genes and the characters that they induce do not segregate in quite the same way as any of the examples that we have considered in the foregoing chapters. Those examples were all autosomal. The easiest way to learn how sex-linked traits are transmitted is to see what happens when a female carrying some recessive sex-linked gene is mated to a normal male, which does not carry that same gene. A very nice example, hemophilia, is available for that purpose, and it is recommended particularly to anyone who teaches genetics to prospective veterinarians.[2]

Sex-linked Hemophilia A

Dogs afflicted with this disease lack the activity of a substance in the blood (antihemophilic factor) that normally interacts with blood platelets (a special kind of blood cell) to facilitate the complicated process of clotting. It is known to specialists studying blood as *factor VIII,* or the *antihemophilic factor* (AHF). There are other kinds of inherited bleeding disorders in mammals, some of which are not sex-linked, but the most common type, which was known for many years in man before it was found in dogs, horses, and cats, is caused in these species by a deficiency of factor VIII activity. It is sometimes referred to as *hemophilia A,* to distinguish it from another kind called *hemophilia B,* of which more anon.

Some affected pups may be born dead, bleed from the umbilicus, or die suddenly in the first few weeks. Others show at about six weeks or later large swellings under the skin (Figure 4-1), lameness, and, eventually, paralysis of one or more legs. The lameness is caused by bleeding into the joints where masses of blood make any movement of the limbs very painful. Bruises experienced during normal play can be fatal to hemophilic pups. One died from the hemorrhage following eruption of a tooth; another died from rupture of a blood vessel when the pup was picked up by its owner (Hutt et al., 1948). In laboratory tests, small samples of blood from hemophilic dogs took 22 to 40 minutes to clot, those times being far above the range of 2.8 to 6.3 minutes for normal dogs (Field et al., 1946).

In dogs (as in man), the deficiency of factor VIII is caused by a recessive, sex-linked gene to which the symbol h has been assigned. The normal allele is H, so that the genotypes and phenotypes possible are as follows:

Class	Genotype	Sex	Phenotype
1	X^HX^H	♀	normal
2	X^HX^h	♀	normal, but carrying h
3	X^HY	♂	normal
4	X^hY	♂	hemophilic

[2]White eyes in fruit flies, long the standard example in many texts, had only a soporific effect on my students. When hemophilia was found in dogs, that problem vanished.

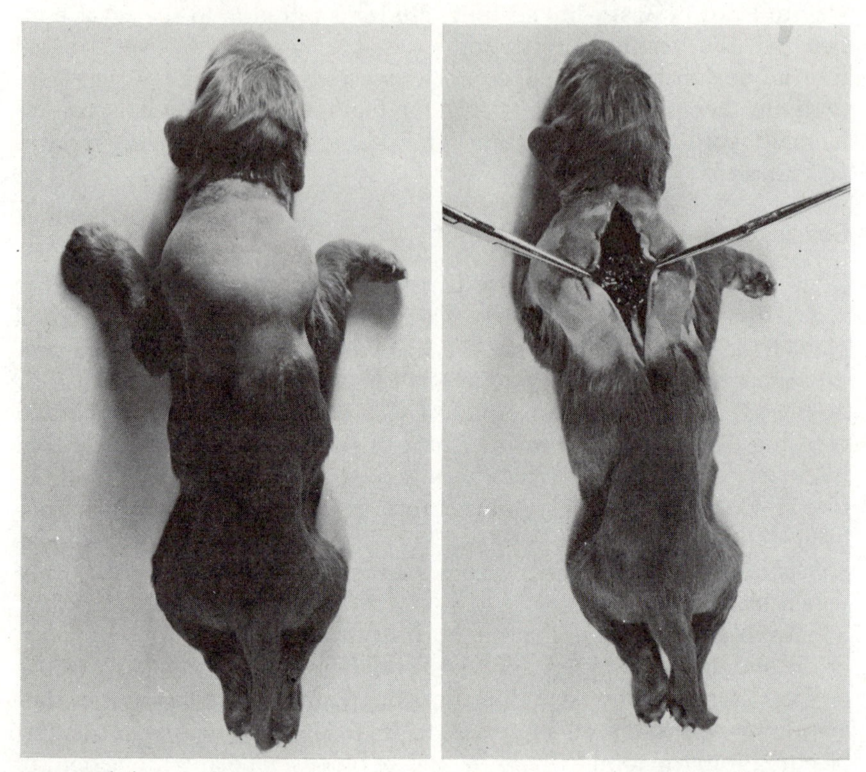

FIGURE 4-1
(a) Hemophilic puppy shaved to show large swollen area under the skin caused by internal bleeding. (b) On dissection, the swelling was found to be full of blood. [From Hutt et al., 1948.]

The first two of these four classes are perfectly healthy animals. Distinction between them can be made only by breeding tests or (sometimes) by laboratory tests for factor VIII in their blood. Both survive equally well. Whenever a male pup is found to have hemophilia A, his dam is thus proven to be a carrier of *h* (class 2 in the table). When such a bitch is mated with any normal male, the expectations in her progeny are as shown in Figure 4-2.

The diagram in that figure illustrates the kind of mating that brings to light any recessive sex-linked gene. It also tells us how to proceed to eliminate such a gene with the least possible loss of otherwise desirable animals. As the bottom line shows, all the females will be phenotypically normal, but the chance that any one of them carries *h* is 1 in 2. Only test-matings or specialized blood tests would distinguish those that do from the ones that do not. One's first thought might be to discard the whole litter, but that is not necessary. The normal brother of a hemophilic pup must have *H* and cannot possibly carry *h*. It is perfectly safe to breed from him.

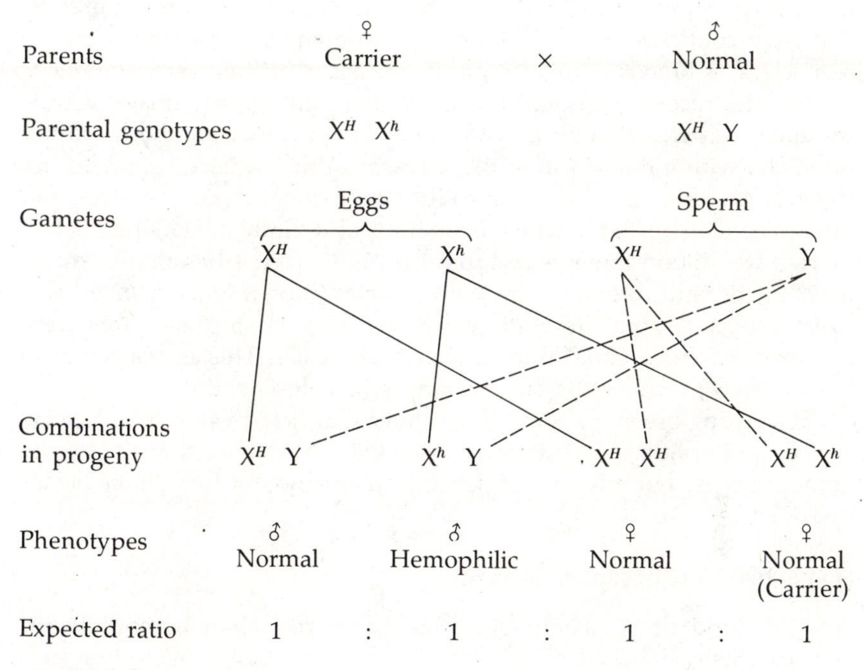

FIGURE 4-2
Transmission of hemophilia A by a carrier female mated to a normal male. X^h = sex chromosome carrying gene for hemophilia; X^H = sex chromosome carrying the normal allele.

It is different with the sisters. If they are otherwise so valuable that the owner is willing to undertake breeding tests to find one or more that do not carry *h*, they can be bred to any normal male—even a mongrel. A single hemophilic son is enough to show that the mother is *Hh*, but, if she is *HH*, it will take 10 or more normal male offspring to be reasonably certain that the eleventh will not be hemophilic. (For more about these probabilities, see Chapter 6.) When it becomes possible to identify the two genotypes with a high degree of accuracy by blood tests for factor VIII, this more practical alternative to the slower test-matings will undoubtedly be used.

A simple way to view all this is that the gene *h* induces the disease in males when it has no opposition. Conversely, the normal allele, *H*, effectively prevents *h* from doing its worst, and is thus a gene protecting the heterozygous female, *Hh*, from hemophilia. It is a gene for resistance to a disease.

Fortunately, *h* is a comparatively rare mutation, but hemophilia A has been reported in many breeds of dogs since 1938 when the Dutch veterinarian Merkens (1938) found it in Greyhounds in the Dutch East Indies. It is now known to have occurred in almost every pure breed and in mongrels. The same mutation is likely to continue to occur in the future. All of this is no reflection whatever on any of these breeds. It merely confirms what has long been known, that mutations are no respecters of breeds or persons and can occur in the best of families.

For the record, it should be stated that hemophilic females can be produced experimentally, and that they occur (albeit rarely) in nature in breeds with a mild form of the disease. In this latter case, males not severely affected survive to reproduce with their carrier relatives and thus produce affected females. Brinkhous and Graham (1950) managed by repeated blood transfusions to raise some of their hemophilic males to maturity, and to mate them with females known to carry *h*. As was to be expected from the mating (h♂ × Hh♀), hemophilic pups were obtained in about equal numbers in both sexes. This is not likely to happen except under the circumstances just described.

If any dog breeder should find a female afflicted with what appears to be hemophilia, investigation is likely to show that it is not hemophilia A, but any one of several other kinds of hereditary bleeding disorders.

Hemophilia B (Christmas Disease)

Another kind of persistent bleeding, more rare than hemophilia A, was originally labelled *Christmas disease* because it was found in a human family whose surname was Christmas. It is now usually called *hemophilia B,* or deficiency of factor IX, a substance in the blood that is necessary for the formation of plasma thromboplastin. It retards the process of clotting, but interferes with it at a point different from that affected by hemophilia A. As in man, hemophilia B is caused in dogs by a sex-linked recessive gene, but it is a different gene from that causing hemophilia A. It has been recognized thus far only in five breeds: Cairn Terriers, Black and Tan Coonhounds, St. Bernards, Cocker Spaniels, and French Bulldogs (Dodds, 1976).

In contrast to hemophilia A, which occurs in severe, moderate and mild forms, all the known cases of hemophilia B have been severe. For both types of hemophilia, however, the incidence and severity of bleeding episodes increase significantly in affected dogs of the larger breeds (Dodds, 1974). Rowsell et al. (1960), who studied hemophilia B in Cairn Terriers, found that one of the afflicted dogs apparently had no trouble until he got into a fight at two years of age. Two others experienced excessive hemorrhage after minor surgery when they were four or five months old. Until their first report of this condition, Rowsell et al. had found no affected females, but in later work, when surviving males were mated to females carrying the causative gene, several hemophilic daughters were obtained (Figure 4-3). Hemophilia B is severe, sometimes fatal, in young pups of St. Bernards and Black and Tan Coonhounds (Dodds, personal communication, 1976).

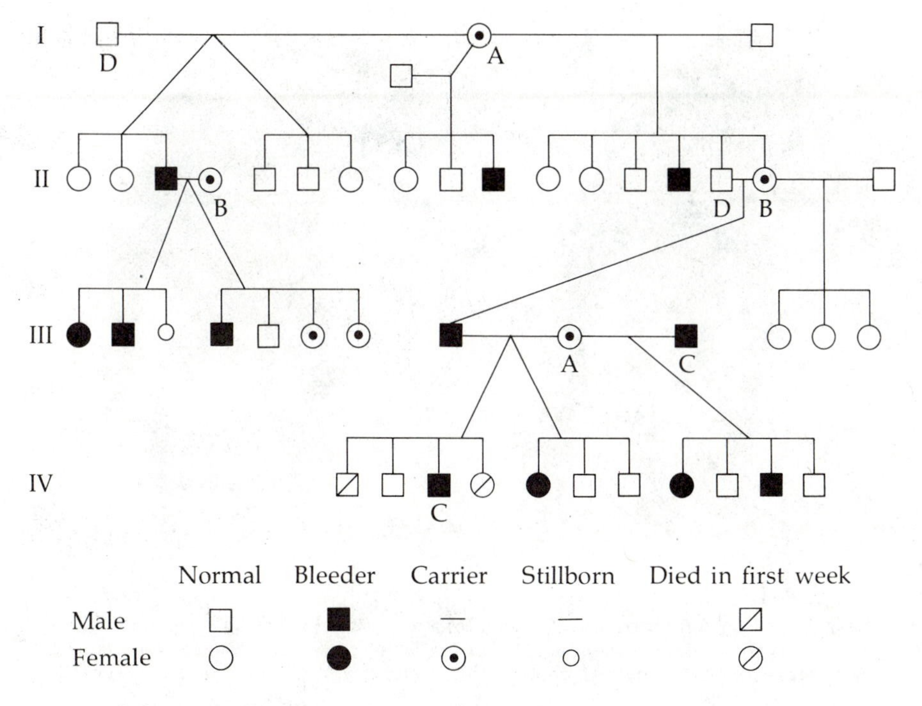

FIGURE 4-3
Hemophilia B (Christmas disease) in Cairn Terriers showing sex-linked, recessive inheritance. Four dogs—A, B, C, and D—all appear twice in the pedigree, and three hemophilic females resulted from mating hemophilic males to carrier females. [Courtesy of H. C. Rowsell.]

Subluxation of the Carpus

For simplicity, let us just call this abnormality *flat front feet.* It appeared in a colony of dogs maintained for studies of hemophilia at the University of North Carolina, and was found by Pick et al. (1967) to be a recessive, sex-linked character. Pups that subsequently showed it appeared to be normal at birth, but at about three weeks, when the pups began to walk, a gradual ventral displacement of the carpal (wrist) bones began. The degrees of final dislocation varied, but in many of the adult dogs it was complete (Figure 4-4).

Search for some anatomical or biochemical basis for the condition was unavailing. Because none of the flat-footers became hemophilic and because no hemophilic dogs had flat feet (Figure 4-5), it seemed probable that the two causative genes were on different X chromosomes in females that carried both. A male could have the gene for one defect, or for the other, but (ordinarily) not both. As we shall see in the next section, breeding tests confirmed that hypothesis.

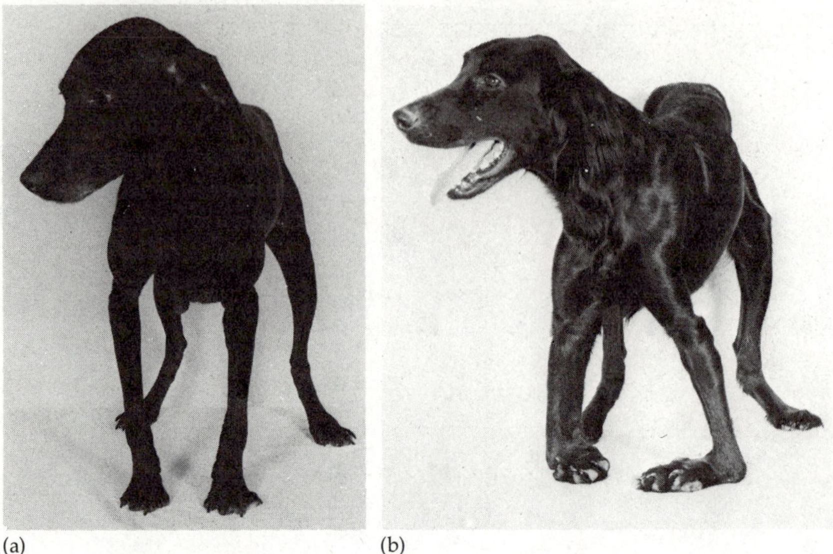

FIGURE 4-4
Normal adult dog (a) for comparison with (b) a male showing flat front feet for subluxation of the carpus. [From Pick et al., 1967. Copyright © 1967, U.S.-Canadian Division of the International Academy of Pathology, the Williams and Wilkins Company, agent.]

Linkage

Because the number of genes comprising the inheritance in any individual far exceeds the number of chromosomes, it is inevitable that any one chromosome must carry many genes. Characters induced by genes all in any one chromosome should, therefore, tend to stick together (i.e., to be associated, or *linked*) more often than characters that segregate independently. However, as we have just seen, sometimes the opposite occurs, and two genes known to be in one chromosome seem to repel each other, so that both do not (or not often) go to one individual.

The answer to this riddle is simple. All the chromosomes occur in pairs. A mutation in one member of a pair does not affect the other member. Thus, many of the dogs at the North Carolina laboratory were there because they carried in one X chromosome the mutation from *H* to *h* that had originally happened in one of their ancestors. Subsequently another mutation occurred in another *X* chromosome, and it induced subluxation of the carpus. For purposes of explanation, let us assign to it the symbol *f*, so that the normal allele must be *F*. (Geneticists have rules for assigning symbols to genes, which are concerned *inter alia* with priorities, but an acceptable list of symbols has not yet been made for the dog's genes. If flat front feet do not merit an *f*, what does?) It is clear that *h* occurred in one X chromosome and *f* in the

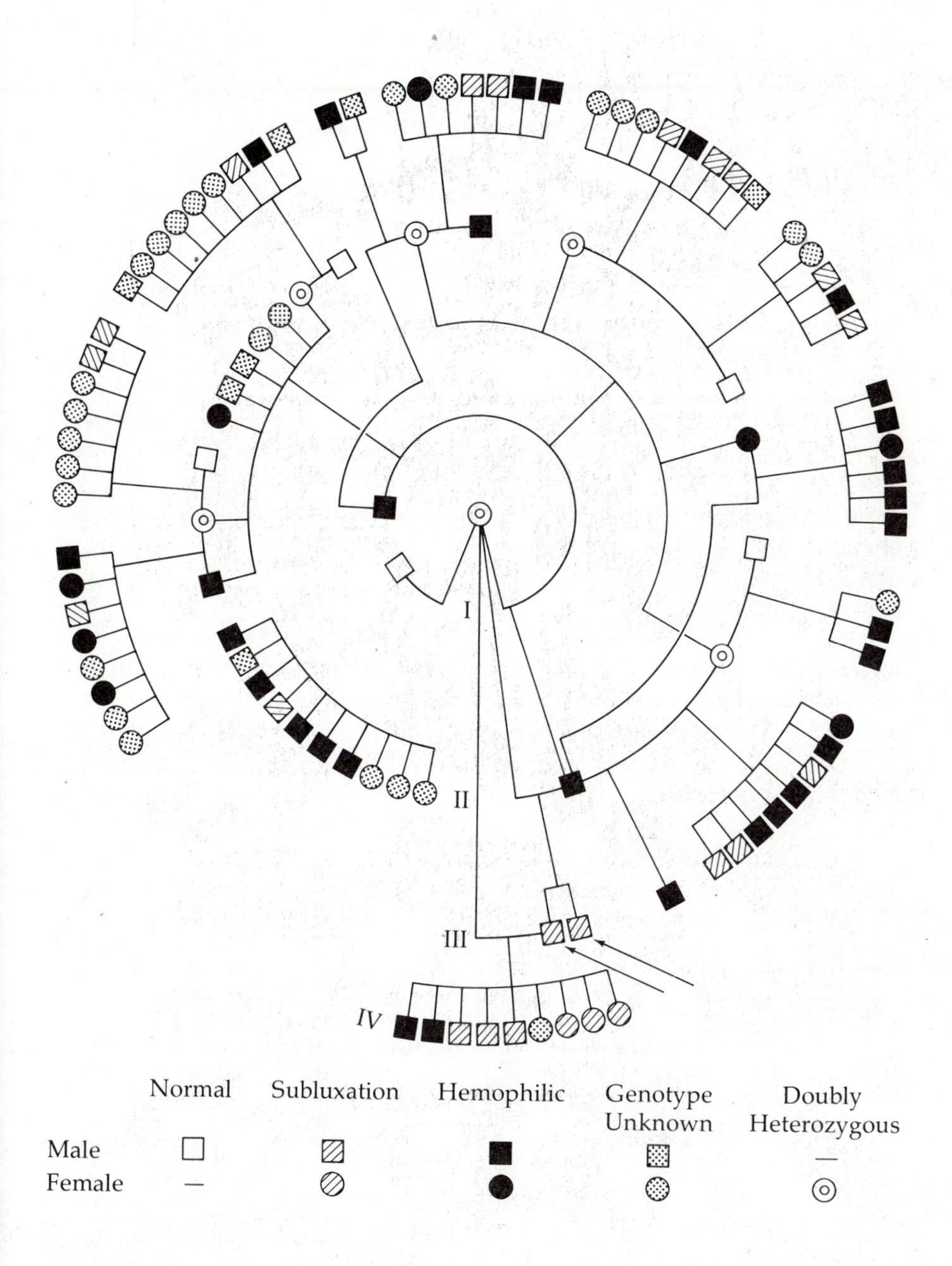

FIGURE 4-5
Pedigree of dogs having subluxation of the carpus, or hemophilia, but never both in one dog. The arrows indicate the two dogs first found to have subluxation of the carpus. For those with genotype unknown, the records were incomplete (males) or undetermined by breeding tests (females). [After Pick et al., 1967. Copyright © 1967, U.S.-Canadian Division of the International Academy of Pathology, the Williams and Wilkins Company, agent.]

other. A phenotypically normal female could get *h* from one parent and *f* from the other. In Figure 4-5, there are five females that did so. Their genotypes were *Hh Ff,* but, to show that the recessive genes were in different chromosomes, they should be written:

$$\frac{H\ f}{h\ F}$$

It is now clear why sons of such females, which could get only one of the maternal X chromosomes, could have flat feet, or hemophilia, but not both.

Linked genes, that both enter any cross (mating) from one parent only, and hence must both lie in the same chromosome, are said to be in the *coupling phase* of linkage when both are dominant, or both recessive. Being in the same chromosome (of the pair), they tend to stick together in the progeny. In contrast, linked genes that come to an individual—one from the sire and the other from the dam—can have one allele in one chromosome of a pair but the other allele in its partner. They are then in the *repulsion phase* of linkage. This is what happened with *h* and *f* at Chapel Hill (Pick et al., 1967).

When gametes are formed, there is usually some interchange of material between partners of a pair of chromosomes. If that had happened in these dogs, most of the female gametes would have been *Hf* or *hF,* but some might have become *hf* or *HF.* Such *crossover gametes* could have resulted in crossover sons with *both* flat feet and hemophilia, or neither. The proportion of such crossovers is very small when the linked genes lie close together in the chromosome, but it increases as does the distance between them. By finding the proportion of crossovers, geneticists can determine which genes are far apart and which are relatively close together. Eventually, *chromosome maps* result.

In this case, Pick and his associates had 49 dogs in which crossovers could have been recognized if they had occurred. Not one was found. Clearly, *H* and *f* are very closely linked, so are *h* and *F.*

Results were different when a study was made by Brinkhous et al. (1973) at Chapel Hill, N.C., in collaboration with Dodds's group at Albany, N.Y., to measure the linkage relations of the two genes causing hemophilia A and hemophilia B. Both were known to be in the X chromosome of the dog and of man. To do this it was necessary to cross two strains of dogs, one carrying the gene for hemophilia A and the other known to carry the gene for hemophilia B. Reciprocal crosses were made (A♂ × B♀, and B♂ × A♀), with four different breeds contributing the mutant genes causing hemophilia A or B. Eventually females carrying both genes (in different X chromosomes) were produced, and these were used in appropriate matings to yield progeny in which crossovers could be identified.

Among 78 of these, in 15 litters, the number of such crossovers was 41 or 52.6 percent. These results were the same as were to be expected if the two genes had been independent; that is, on different chromosomes. How can they be reconciled with the previous, conclusive evidence that the two genes were on the same chromosome? Very simply. From linkage in other species (particularly that useful fruit fly, Drosophila), it is known that, when crossovers between genes in one chromosome exceed 50 percent, the two genes are too far apart to show any linkage. Accordingly, the true *map distance* between these two genes will not be known until each can be tested for linkage with other sex-linked genes, so that crossovers can be measured between genes closer together. It is of interest that estimates also place the sex-linked genes for hemophilias A and B far apart in man.[3]

Although the mutant genes causing hemophilias A and B lie far apart in the dog's X chromosome, those two disorders can be associated in another way. Slappendel (1975) found that 10 male French Bulldogs sent independently from various parts of the Netherlands to a clinic in Utrecht all had hemophilia and were all descended (through females) from a common ancestor, Marion. Oddly enough, blood tests for eight of them revealed that the 5 in one line of descent all had hemophilia A, but the 3 in another line all had hemophilia B. The most likely interpretation was that Marion had been doubly heterozygous, with the gene for A in one X chromosome, the gene for B in the other, and protection against ill effects of either provided by their normal alleles. That is exactly the genotype produced experimentally by Brinkhous et al. (1973) for their linkage studies.

Other Sex-linked Genes

In addition to the three sex-linked mutations discussed in this chapter, only one other is known so far in dogs. It is the muscular dystrophy found in Irish Terriers by Wentink et al. (1972) in the Netherlands. Details are given in Chapter 9.

It is reasonably certain that others will eventually be found. In male mammals, any mutation in the X chromosome will not be hidden by a protective allele in that chromosome's partner because the Y chromosome carries few genes if any. A recessive mutation in an autosome can be carried along unseen for several generations before chance brings two carriers together. However, if a recessive mutation occurs in an X chromosome of a female mammal, she will not show it, but about half of her sons will do so. Any defect that shows up in one sex but not in

---

[3]Relax! In man, they do it from studies of the rare pedigrees in which hemophilia (A or B) is associated with certain kinds of color-blindness or other traits known to be sex-linked.

the other is easily recognized as being probably sex-linked. It is not surprising that many such conditions are known in man, or that for the fowl (in which the female is heterogametic) almost as many genes have been spotted in the sex chromosome as in the other linkage groups (each presumably representing one chromosome) that are known.

Many mutations are lethal to homozygotes and to individuals that are *hemizygous*; that is, have only one sex chromosome that carries genes. For some years, it was thought that prenatal mortality in dogs, pigs, man, and other mammals fell more heavily on males than on females because the former were not protected against sex-linked lethal genes, whereas the females could carry protective alleles in their other sex chromosome. This theory was proven untenable when it was found that for moths and birds (in which the females have only one sex chromosome) mortality rates were also higher in males than in females.

Reciprocal Crosses

Reciprocal crosses were mentioned earlier, but, as they are particularly useful in studying sex-linked inheritance, further explanation is in order. Reciprocal crosses between red and black dogs would be, for example, red ♂ × black ♀, and black ♂ × red ♀. Because the genes responsible for these colors are autosomal, the F_1 generations from both crosses would be the same—black.

Things are different if reciprocal crosses are made when some sex-linked character is involved. With tender, loving care, it is now possible to keep hemophilic dogs alive for several years and thus to mate hemophilic animals of both sexes. If reciprocal crosses are made between hemophilic and normal dogs, the results in the F_1 generations differ markedly, depending on which parent was hemophilic. They differ also in the F_2 generation (Figure 4-6).

Because of these differences, reciprocal crosses provide a useful test to determine whether some trait is sex-linked or autosomal. They need not be carried to F_2 generations when the numbers of F_1 animals are adequate to prevent faulty conclusions from variations by chance in small numbers. Reciprocal crosses are also useful to detect non-genetic influences such as the relation between size of the mother and size of her offspring. They are commonly used by animal breeders when breeds or strains are crossed in the quest for hybrid vigor, because they provide better sampling of the breeds than when each breed or strain is represented by one sex only.

Sex-limited Traits

It is clear that if any genetic character in mammals were caused by a gene on the Y chromosome, only males could show it. It would be

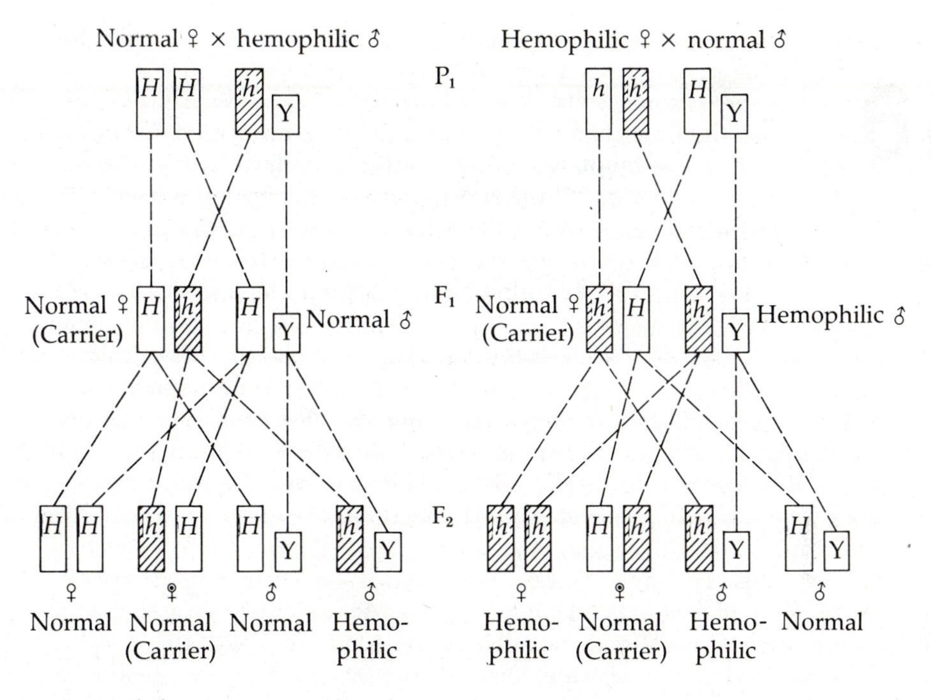

FIGURE 4-6
Diagrams showing the differing results in reciprocal crosses when a sex-linked recessive character (hemophilia) is crossed with its normal allele.

sex-limited. One supposed such trait in man was carried for years in the textbooks until good reasons for doubting the accuracy of the pedigree emerged from a reinvestigation of it. No Y-borne genes are known in the dog or in any other mammal except ♂-determiners.

Other genetic variations are truly sex-limited because they can be manifested by only one sex. Familiar examples include genetic differences among cows in ability to produce milk, and genetic differences among hens in egg production, size of egg, and color of egg-shell. Males carry genes that influence these things, but in that sex the genotype can be revealed only by the performance of the daughters.

Cryptorchidism

This defect appeared at one time to provide a perfect example in the dog of a sex-limited hereditary condition. It may still do so, but there is now some question about the extent to which it is genetic in origin. As it is a problem for many a breeder, some discussion of it is warranted.

Cryptorchidism is the failure of the testes (one or both) to descend into the scrotum through the inguinal canal. When only one testis is retained within the body cavity and the other descends normally, some people refer to that condition as "monorchid" (one testis), but that term is a misnomer. Cryptorchids are undesirable, not only because they are disqualified in the show-ring (in the United States) but also because the undescended testis becomes tumorous and the dog is feminized. Bilateral cryptorchids are sterile, but those with only one retained testis are fertile. The condition occurs in other mammals.

Theories about a genetic basis for cryptorchidism date back to a report by Härtl (1938) of 57 litters of Boxers, among which 23 percent of the 168 males were cryptorchids. In his 55 litters having at least one affected male, the incidence in all males was 36 percent. Härtl believed the defect to be a simple recessive trait, sex-limited, but not sex-linked. He traced the causative gene back to four foundation sires, all related. There are several other reports suggesting that there must be a genetic basis, and that cryptorchidism is probably a simple recessive defect. It seems unnecessary to review all those reports here, because they do not provide conclusive evidence. The fact that they all find it recessive could mean only that knowledgeable breeders prefer to breed from normal dogs, and not—if they can avoid it—from unilateral cryptorchids. Mendelian ratios adequate to prove that a single recessive gene (when homozygous) is responsible are not available. Genetic analyses are complicated because females are unaffected and their genotypes can only be assumed.

Some years ago, when evidence seemed to be accumulating to show that cryptorchidism is a simple recessive defect, I published a table giving results to be expected in six kinds of matings involving that condition (Hutt, 1964). It now seems possible that it is influenced by a number of genes, and probably even by influences not genetic. Some of the genes involved may induce the defect by influencing hormones or muscles. Proof of any of these theories will await further research. Meanwhile, readers are referred to good reviews of the problem by Willis (1963) and by Burns and Fraser (1966).

It is interesting to note how the attitudes of kennel clubs toward cryptorchidism have varied. Willis (1963) states that in 1958 and 1959 the German breeders of German Shepherds published the names of 111 unilateral and 14 bilateral cryptorchids whose progeny would not be registered. In 1955, the British Veterinary Association (1955) warned dog breeders not to breed from any cryptorchids, their litter-mates, their parents, or any normal males sired by (unilateral) cryptorchids. More recently, the Kennel Club in Britain has withdrawn former restrictions and does not exclude cryptorchids from breeding. The American Kennel Club will register cryptorchids, in spite of the fact that they are disqualified in the show-ring.

These differing viewpoints are confusing. If there is any genetic predisposition to cryptorchidism, breeders should avoid using the fertile, unilateral cryptorchids and their close relatives. What is really needed now is not any further search for Mendelian ratios, but an experiment in which a deliberate attempt is made to eliminate the defect by selection against it in one strain of dogs, while, at the same time and under the same environmental conditions, an attempt is made to *increase* it in another strain of the same breed. In such an attempt, the breeders would deliberately use unilateral cryptorchids and females that are close relatives of cryptorchids. Do such males beget more affected dogs than others? If so, no further proof of a genetic predisposition to cryptorchidism would be necessary.

This experiment would probably have to be carried on for several generations to get a conclusive answer to the question. No average breeder with a kennel of average size could afford to do it—but a breed association might arrange for a cooperative experiment with several participating breeders. To do it adequately might entail the use of a score of breeding females in each generation, progeny-tests of many sires, and supervision by a geneticist competent to utilize these progeny-tests to their maximum possible extent.

Until some of these questions are answered, this author would avoid use of cryptorchids for breeding, just as he would do with any other abnormality that might possibly be genetic in origin.

5 Quantitative Characters

We come now to what appears at first glance to be a kind of heredity different from that considered thus far. In most of the examples of genetic traits in the previous chapters, dogs either had them or did not. There was always a clear distinction between the two classes. Puppies having a lethal character (bird-tongue) died; heterozygotes did not show that trait and lived. Black dogs were easily distinguished from red ones. Even the incompletely dominant gene causing the merle pattern permitted easy distinction between the beautiful heterozygotes and the defective homozygotes. Similarly, dogs afflicted with hemophilia A are easily distinguished from those that are not.

In all these cases, the animals showed *discontinuous* variation. They had one thing or the other—not both—and could therefore be easily sorted into two classes, or, when an incompletely dominant gene was operating, into three.

Continuous Variation

In contrast to these comparatively simple cases in which a single pair of alleles seems to determine what the phenotype is to be, there is a kind of inheritance in which many genes act together to induce differing amounts or degrees of one and the same character. They cause *continuous* variation in that character, so that differences among individuals are often scarcely perceptible in a population, flock, or kennel in which all members show it to some degree. Because such gradations are best determined by weighing, by measuring, or by recording productivity, traits dependent on this kind of heredity are labelled by geneticists as *quantitative* or *metric* characters.

Familiar examples of hereditary quantitative characters in man are stature (height), duration of life, and intelligence. These three traits are probably also hereditary in dogs, but they have not been as much studied in dogs as in man. Nearly everything of economic value that domestic animals produce is influenced by quantitative inheritance (and also by the environment), and the breeder who selects animals to yield more milk, meat, eggs, or wool, or to withstand disease, is not concerned with Mendelian ratios. His objective is to accumulate in his stock the genes that tend to produce superior animals, and to eliminate genes that interfere in any way with maximum performance, optimum appearance, or whatever goals he seeks to attain. Of course, the breeder doesn't think in terms of genes, nor does he need to do so. He selects toward his ideals. The genetic part of the variation that encourages him to continue selection depends on the interaction of many genes—too numerous to be counted.

Polygenic Inheritance

In the early days of genetics, when biologists were trying to fit various inherited traits into the Procrustean limits of nice Mendelian ratios, those that refused to do so were commonly said to show "blending inheritance." That expression was used, because, when the usual Mendelian analysis was attempted and crosses were made between contrasting types (big × little, tall × short, and so on), the F_1 generation showed no dominance of either parental type, but a blend of both. It might be closer to one parent than to the other, but it was intermediate between the two.

Eventually, the blending was proven to result from the interaction of many genes. For some years thereafter, while units of inheritance were still called *factors,* such inheritance was called *multifactorial,* but when the Latin *factors* became Greek *genes,* a corresponding change in the prefix seemed desirable, and genetic characters showing continuous variation are now attributed to *polygenic* inheritance. The word is different, but the music is the same.

With quantitative characters, the range from the least to the greatest may be large or small. Some idea of that range in a population can be had by measuring each individual in it (weight, height, or any other quantitative character), making some 6 to 12 arbitrary classes from least to greatest, and then finding how many individuals fall into one or another of these classes. About half the population should fall below the mean (average) and half above it. If the means for the different classes are then plotted along one axis, and the frequencies in each class on another at a right angle, a line connecting those frequencies will form a *curve of normal distribution.* That curve may be tall and narrow, showing little variation on each side of the mean for the whole population, or it may be lower and wider (Figure 5-1). From such graphic measures of the variation within the population, the breeder can get some idea of how much material is available for selection if he wants to raise or lower the average for the whole population.

It is unnecessary for dog breeders to measure or weigh all their animals or to draw such curves, but it is desirable to know about normal variation in polygenic traits, and how such traits behave in crosses or under the pressures of selection. One lesson evident in Figure 5-1 is that no single individual can represent his population (breed, strain) with respect to some quantitative character unless he lies close to the mean for that character.

When individuals are crossed that differ widely in some polygenic trait, the F_1 generation is likely to be intermediate and fairly uniform. If representative F_1 animals are then intermated, variation in the resultant F_2 generation is usually far greater than in the F_1, and can even result in extremes beyond the figures for the original parents. If the F_1 stock is backcrossed to the smaller, shorter, or lesser parent, the resulting backcross population will be shifted in that direction. Conversely, if the F_1 animals are backcrossed to the larger, taller, or greater parent, their backcross offspring will be grouped around a mean closer to that parent. Distributions of these two backcross populations may overlap, but there should be a distinct difference between their means.

It would be nice if such typical polygenic inheritance could be illustrated with a suitable example in dogs, but, as none is known to me, I

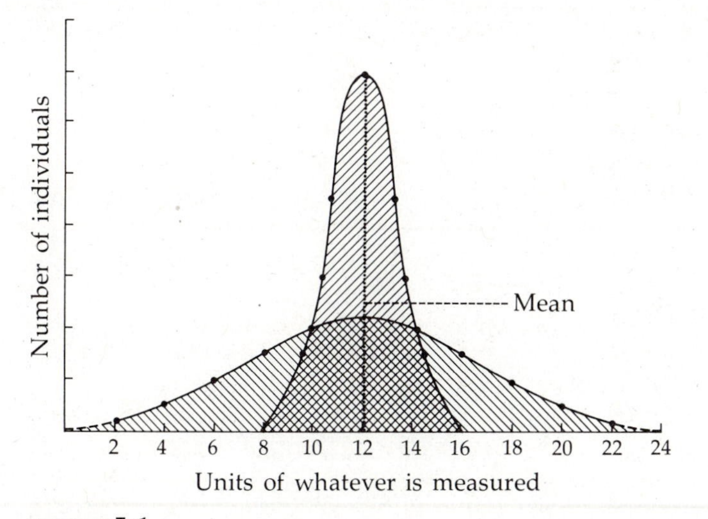

FIGURE 5-1

Idealized curves of normal distribution for two populations having the same mean, but one much more variable than the other and hence more responsive to selection.

use instead some results in my own experiments with a polygenic trait in chickens. In four generations of selection, two lines of White Leghorns were differentiated that differed significantly and consistently in their ability (as chicks) to grow on a diet deficient in the amino acid, arginine. The measure of that ability was their weight at four weeks of age, after being on the deficient diet since hatching. In our laboratory jargon, the strain that could do so very well was labelled the LA or Low-A line, because its requirement of arginine was comparatively low. Conversely, the other stock became the HA or High-A strain (Hutt and Nesheim, 1968).

When these two strains were crossed (using several males and females of each to ensure adequate sampling), they produced an F_1 generation that was intermediate (Figure 5-2). Reciprocal backcrosses of that F_1 generation to the two parent strains yielded backcross populations in which the means were shifted away from that of the F_1 generation and toward those of the original pure-line parents. This illustrates typical behavior of a polygenic trait in such crosses.

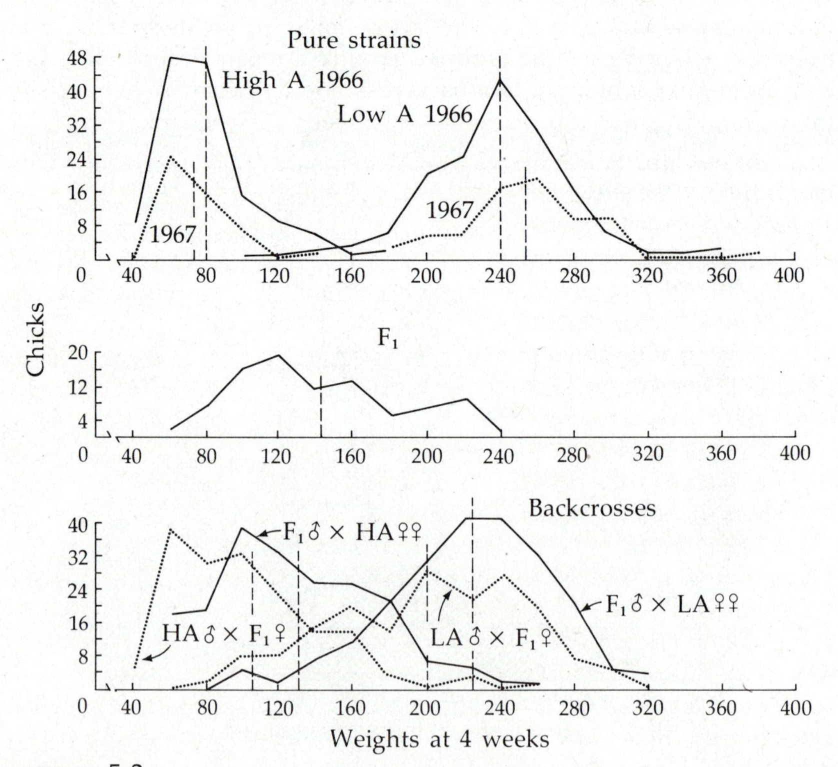

FIGURE 5-2

Polygenic inheritance of variations in the utilization of arginine by chicks. Weights of chicks after four weeks on the deficient diet. Broken vertical lines show means for populations. See details in text. [From Hutt and Nesheim, 1968.]

The example also shows an important point to be remembered about polygenic traits. All of the frequency distributions in Figure 5-2 resemble the normal curves shown earlier, but do so only roughly. They are not smooth. Those in Figure 5-1 are ideal curves, based on large, theoretical populations. For those of Figure 5-2, the number of chicks per group varied from 49 to 154. Had there been 300 or more in each, the corresponding frequency distributions would have fitted more closely the ideal normal curves expected for any polygenic trait in a large population. A single litter does not adequately represent the normal variation in a breed with respect to some quantitative character such as size, height, or intelligence. Fifty dogs of one breed at a large show would do it better.

Environmental Modification of Polygenic Traits

Practically all polygenic traits can be influenced in some degree by various factors in the environment. For example, body weight (in man and dogs) can be drastically reduced by disease or by semi-starvation and can be increased by overeating and lack of exercise. As a measure of size, it is far less desirable than height, which is more dependent on skeletal measurements. Growth of the skeleton can be stunted by adverse environmental conditions before maturity, but is affected less, or not at all, thereafter.

A puppy having at birth four generations of champions behind him is a potential champion too, but whatever good qualities have been handed down by his illustrious ancestors can be completely suppressed if the puppy is run over by an automobile at three months. The automobile is an inescapable part of the environment.

Animals differ genetically in ability to resist disease (Hutt, 1958). In most cases, that ability depends on the interaction of many genes, but it also depends on the various factors that make up the environment. Resistance at comfortable ambient temperatures can be entirely lost if the animal is badly chilled, or may be partially broken down if the chilling is less severe. It varies also with the severity of exposure to the organisms causing disease. Animals that are genetically susceptible to some disease will not reveal that weakness if they are never exposed to the pathogen that causes it. Similarly, as every dog owner knows, vaccination against distemper provides a favorable environment for resisting that disease, even in dogs that might otherwise be genetically susceptible to it.

It is unnecessary to belabor the fact that food is an important part of the environment. Genetic ability to produce milk, meat, eggs, or other desirable animal products is best revealed on optimum diets and can be obscured or completely suppressed on those poorly balanced or not provided in adequate amounts.

In the average dog (or person), the appearance, conformation, and performance are far more dependent on polygenic inheritance than on single genes that segregate nicely in Mendelian ratios. It should not be necessary to list further examples of environmental influences to convince the reader that the degree to which any polygenic trait is expressed depends not only on the genes with which the animal is born, but also on the conditions under which those genes subsequently have to operate.

In days gone by, when debates were more fashionable than they are now, a favorite topic was "Is heredity more important than environment?" Most geneticists would now agree that, while there are situations and conditions in which one of these two forces appears to be more important than the other, so far as quantitative traits are concerned, the outcome depends on the interaction of both.

Moreover, we have to recognize that some single genes are lethal when homozygous, that others cause severe abnormalities, and that, in these and similar cases, the environment has little or nothing to do with the outcome. Conversely, many an animal is killed by an adverse environment regardless of what genes it carries.

Size and Conformation

To shed some light on differences among breeds of dogs in size and conformation, Stockard (1941) made a number of crosses between breeds that differed greatly in these respects. From reciprocal crosses between Basset Hounds and German Shepherds, the F_1 progeny were intermediate in size between the two parents, with legs tending to be short like those of the Basset Hound. Among 144 dogs of the F_2 generation about three-quarters had "typical achondroplasic short legs," and only a quarter had long legs like those of the P_1 German Shepherd. From these results and a very limited backcross, Stockard concluded that length of leg in the dog is determined by a single pair of genes. Some geneticists (including myself) would say that further evidence on that point is desirable. It is not clear that the 144 dogs of the F_2 generation fell into two or (at most) three distinct classes as they should have done if length of leg depended on a single pair of alleles. Most measurements of size show continuous variation from one extreme to the other.

That same population showed nicely how wide a range can occur in the F_2 generation from parents contrasting greatly in some quantitative character. Even in a single litter of seven dogs there were extreme differences in height and size (Figure 5-3).

The many genes that influence the expression of some polygenic trait do not necessarily all exert equal effects on that trait. Some may enhance it, others reduce it. It is the variability among such genes that

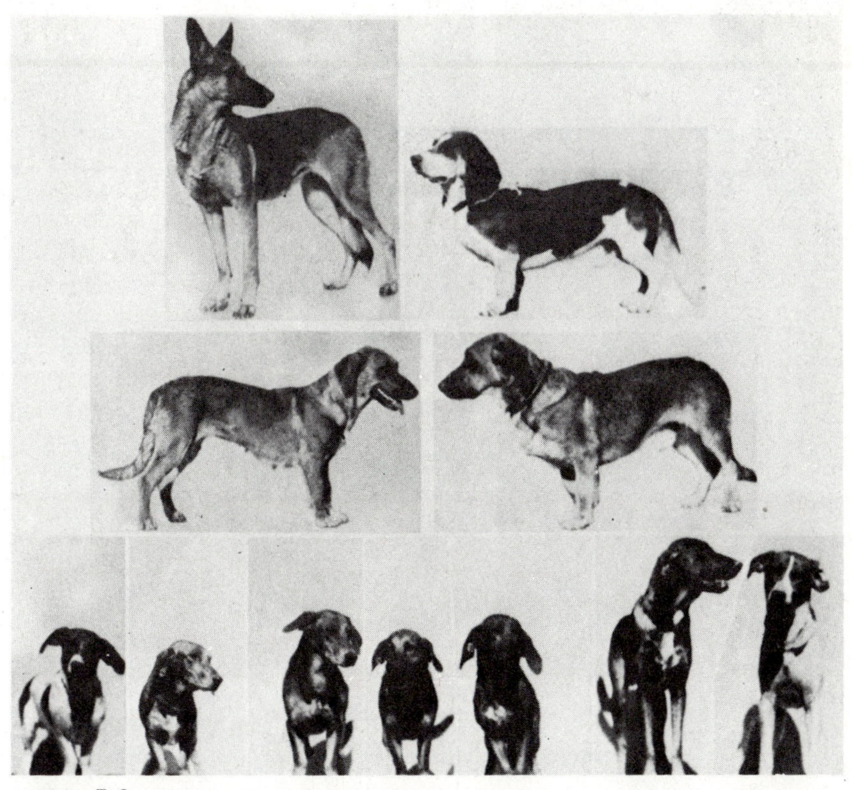

FIGURE 5-3
Cross of German Shepherd ♀ × Basset Hound ♂ showing short legs and intermediate size in the F_1 generation, but, in a single F_2 litter of seven, extreme variation in size. [From Stockard, 1941, p. 51. Courtesy of Wistar Institute, Philadelphia.]

provides the material for selection. Put more simply, it allows the breeder to select for bigger dogs or smaller ones as he prefers.

Apart from polygenic inheritance, in some species single genes have been identified which exert an altogether disproportionate effect on size. One of these, in the fowl, is a sex-linked recessive gene that reduces body size by about 30 percent in females (which are hemizygous) and by about 42 percent in homozygous males. In the same species, an autosomal gene reduces size in a lesser degree but segregates clearly, leaving the dwarfs quite distinct from their normal siblings. Hereditary dwarfism caused by a single gene occurs in the dog, and those affected have an abnormality of an endocrine gland (see Chapter 12 and Figure 12-2).

It would be interesting to find out to what extent (if any) the smaller breeds of dogs have been made small by single genes that (as in the examples just given) override the usual determination of body size by a complex of polygenes.

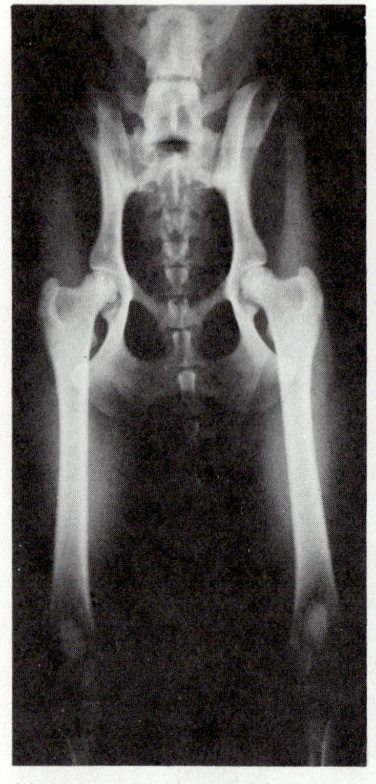

(a)

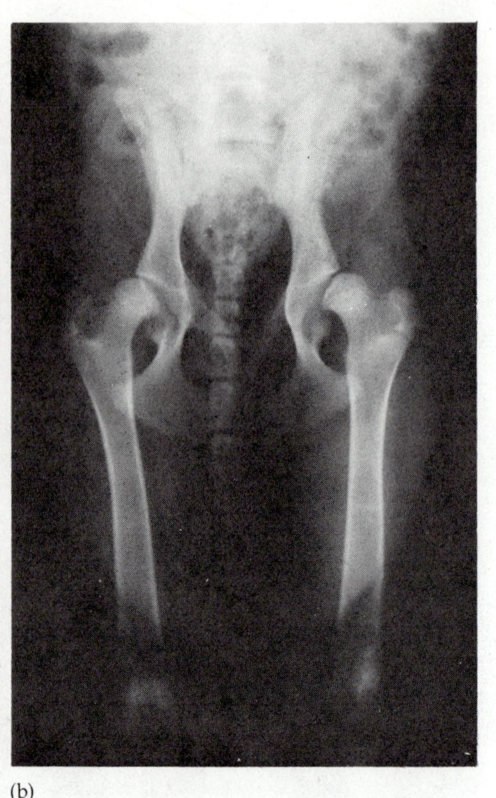

(b)

FIGURE 5-4
Radiographs of hip joints of Labrador Retrievers. (a) Normal, at two years of age. (b) A dog with moderate to severe bilateral hip dysplasia at seven months. Note (1) subluxation of the head of the femur, (2) that it is flattened, and (3) shallowness of the acetabulum. [Courtesy of George Lust, James A. Baker Institute for Animal Health, Cornell University.]

Finally, body size in dogs is limited to some extent by the size of the dam. Manchester Terrier bitches will not produce offspring that become as big as Great Danes. The same maternal limitation of ultimate size is found in other mammals right up to horses and ponies. That limitation is not caused by the mother's genes. It is a restriction, imposed by the prenatal environment, which cannot be entirely overcome after the animal is born.

Hip Dysplasia

In the dog, hip dysplasia provides a good example of a defect that is polygenic. It has been so much discussed, described, and deplored in the last two or three decades that a detailed description should be unnecessary here. Let us just say that it is a failure of the head of the femur to fit snugly into the acetabulum, with resulting degrees of lameness and faulty gait that vary greatly (Figure 5-4). The degrees of

TABLE 5-1 The Incidence of Hip Dysplasia in Progeny from Three Types of Matings with German Shepherds of the Swedish Army

Type of Parents	Progeny	
	Number	Dysplastic, Percent
Series 1 Both free of dysplasia	462	37.5
Series 2		
(a) Normal ♂ × dysplastic ♀	213	41.3
(b) Dysplastic ♂ × normal ♀	89	52.8
(a + b) combined	302	44.7
Series 3 Both with dysplastic hips	88	84.1

Source: Data of Henricson et al. (1966).

subluxation of the femur also vary. They can be recognized fairly accurately by good examination with X rays, and it is now customary for specialists in such examinations to classify the degree of dysplasia as falling into one of four or five somewhat arbitrary grades of subluxation. Sometimes it is unilateral, particularly in the less severe cases.

Like many another quantitative character, when the influence of heredity on it was recognized, hip dysplasia was at first said to be a simple recessive trait, or to be a dominant one with incomplete penetrance. It is now recognized that the condition results from the action of an undetermined number of genes; i.e., it is polygenic (Hutt, 1967).

Conclusive evidence of the genetic nature of this defect came from studies of its frequency in German Shepherds[1] of the Swedish Army. Diagnoses were made by X-ray examination between one and two years of age. The frequencies of dysplastic offspring from matings of different kinds, as given by Henricson et al. (1966), are shown in Table 5-1.

The incidence of dysplastic progeny from parents both dysplastic was more than twice the frequency in dogs from parents both normal. When only one parent was affected (Series 2) the proportion of dysplastic progeny (44.7 percent) was intermediate between that in Series 1 and 3. The difference between the reciprocal crosses of Series 2 is probably not significant.

These three kinds of matings show that predisposition to hip dysplasia is hereditary, and they also illustrate fairly well how polygenic traits behave in crosses in which they are involved. It is at first glance surprising that the dogs of Series 2, although intermediate, had only 44.7 percent affected; i.e., as a group they resembled their

[1]To avoid any possible ambiguity from following common usage among dog breeders, and to placate a friendly critic who spotted 28 places in which I have not accorded full honors to German Shepherd Dogs, it must be made clear that all the German Shepherds mentioned in this book are dogs—not German shepherds.

normal parent more than the dysplastic one. It seems possible that some of the dysplastic parents in Series 2 may have had lower grades of hip dysplasia than those in Series 3.

An important lesson to be learned from Table 5-1 is that—contrary to the old adage—like does not always beget like, particularly when quantitative characters are concerned. As the geneticist would put it, the phenotype does not always reveal the genotype; the progeny tell more. In this case, that 37.5 percent incidence of dysplastic progeny from normal parents tells us that genes tending to induce hip dysplasia were widespread in those German Shepherds, and that a radiograph showing normal hips, although highly desirable, is no guarantee of normal hips in the progeny. Similarly, parents that are superior with respect to any quantitative character (champions, perhaps!) are more likely to produce superior progeny than are parents less fortunate, but they cannot be counted on to do so.

Grades of Hip Dysplasia

After it became recognized that heredity had more effect in causing hip dysplasia than had any other known influence, it was not long before the clarion call rang out, "Never breed from a dog with hip dysplasia." While this may have resulted in more dogs being X-rayed than ever before, it also caused dismay in many a breeder who had only one or two sires, both slightly dysplastic, to breed from, and who would have been out of business if he adhered strictly to the injunction. With incidence of the defect running to 50 percent or more in some breeds, many a breeder was in a quandary about what to do. Fortunately, some further data from the German Shepherds of the Swedish Army settled the problem. They provided conclusive answers to the following questions:

1. Is the degree of hip dysplasia in the progeny related to the grade of it in the sire or dam?
2. Should one breed from dogs classified after X-ray examination as borderline or grade 1?

Because the data of Henricson et al. (1966) on these points are more extensive than we are likely to find elsewhere, they are summarized here (Table 5-2).

It is clear that sires and dams showing only grade 1, or borderline, had very little more severe hip dysplasia (grades 2–4) in their progeny than did those with normal hips. The difference was scarcely significant for either sires or dams. Similar results were found in Switzerland by Freudiger (personal communication, 1976).

Far more significant is the evidence that among offspring of a parent showing severe hip dysplasia (grades 2–4) the proportion afflicted

TABLE 5-2 Relations Between Grade of Hip Dysplasia in Progeny to That in the Sire or Dam

Progeny	Grades of Dysplasia in Sire or Dam		
	None	B[a] or 1	2, 3, or 4[b]
		A. In Sire	
Number	682	22	152
Proportion (percent) with dysplasia of grade:			
None	59.4	63.6	24.3
Grade B[a] or 1	22.6	13.6	28.3
Grade 2, 3, or 4	18.0	22.7	47.6
		B. In Dam	
Number	586	131	171
Proportion (percent) with dysplasia of grade:			
None	60.4	57.2	29.2
Grade B[a] or 1	22.9	23.7	25.7
Grade 2, 3, or 4	16.7	19.1	45.0

[a] Borderline.
[b] Only 7 progeny were classified as grade 4, all from one or more dams of that same grade. No sires of grade 4 were used.

Source: Condensed, with conversions to percentages, from data of Henricson et al. (1966) for German Shepherds of the Swedish Army.

to that same degree is more than twice as great as in the progeny of a parent that is normal, borderline, or grade 1.

For the breeder, the message is clear. When a dog is X-rayed, if it is found to have hip dysplasia, that diagnosis is not complete unless the degree of dysplasia is given. On the usual scale of grades 1–4, dogs showing anything worse than grade 1 should not be allowed to reproduce. If the breeder has plenty of animals with normal hips, they should be given preference (other desiderata being equal) over dogs of grade 1. However, if only dogs showing grade 1 are available, they can be used without much more risk of getting badly defective progeny than when only dogs with normal hips are bred. It is well to remember also that, as Table 5-1 shows, it is not so risky to have only one parent dysplastic as it is when both show dysplasia. Breeding to reduce the condition is discussed further in Chapter 14.

Considering the dogs that are normal or borderline as being at one end of a continuous gradation, and those of grade 4 at the other, it is clear that in this case like does *tend* to beget like, but that exceptions occurred at both ends of the distribution. The normal parents produced some dogs with severe dysplasia, and even the most defective parents begot some normal progeny. Similar paradoxical results can be expected with any polygenic trait.

Hip Dysplasia: Unanswered Questions

Considerable study and much discussion have been directed toward answering two questions about hip dysplasia: (1) What causes it? (2) What can the breeder do (apart from selection) to lessen the chances of a dog becoming dysplastic? The first of these is really asking how the genes that induce it exert their effects. The second asks what adverse environmental influences are known, and how they can best be avoided.

More studies have been made on the first question than on the second. The defect has been attributed by various writers to shallowness of the acetabulum, endocrine secretions, laxity of the hip joint, downward slope of the back from withers to hindquarters, spasms or shortening of the pectineus muscle, and other causes. As there seems to be no agreement among proponents of these theories, I beg to be excused from discussing them.

One point on which all have to agree is that hip dysplasia is more frequent in big breeds than in small ones. Furthermore, even within one breed (German Shepherd) Riser et al. (1964) found that, among 222 dogs weighed at 60 days of age, those then heavier than average (6.1 kg) later had a higher proportion with hip dysplasia than did those below average weight at 60 days.

Among Greyhounds, hip dysplasia is comparatively rare. Comparison of their musculature with that of German Shepherds (among which the defect is very common), led Riser and Shirer (1967) to conclude that dogs having greater "pelvic muscle mass" are less likely to become dysplastic than others. Although there is apparently not unanimous agreement on that score, such studies are desirable because they may eventually reveal some anatomical or physiological characteristic that would serve as an indicator of predisposition to hip dysplasia and thus of a type that breeders should avoid.

None of the causes just mentioned gives any clue as to significant environmental influences that might aggravate or abate a genetic predispositon to hip dysplasia. From Riser's evidence that big pups were more likely to become dysplastic than smaller ones, it was thought by some people that a diet not conducive to rapid growth would afford some protection. Experiments to test that idea are now in progress.

Most polygenic traits are influenced to some extent by the environment. For hip dysplasia, that environment could be operative prenatally (i.e., a maternal effect, like that limiting body size), or after birth, or both. It is to be hoped that whatever environmental forces there may be that tend to cause hip dysplasia will eventually be discovered.

Further information about quantitative characters and polygenic inheritance can be found in Hutt (1964).

6 Probabilities and Tests

One does not have to be a mathematician in order to breed champions, but there are times when a little familiarity with simple probabilities may be useful. With increasing knowledge of genetic defects that are caused by single genes, and recognition of the need for breeding tests to help elimination of such defects, it becomes important to know how to make the appropriate tests and how to interpret the results obtained from them.

Test-crosses with a Viable Homozygote

A test-cross is the mating of an animal of unknown genotype (A) to one or more of known genotype (B) to find out whether or not A carries some specific recessive gene or genes. The simplest probability with which to introduce the kinds of tests that a dog breeder might wish to make is that known to some of us as the 50 : 50 chance, to others as 1 in 2, or ½, and to biometricians as $p = 0.5$. To see how it works, let us suppose that some breeder of Cocker Spaniels is upset because one of his best black stud sires has thrown red puppies from a black bitch. That is not the way to win friends and influence people. Nor does it matter that the black bitch must have been equally to blame (see Table 2-1).

It is a fairly simple procedure to test black sires and thus to ensure that any which could throw red offspring are identified. As we have seen in Table 2-1, the safest sire (for those who want no red or brown pups) should have the genotype *BB EE*. To explain the test-cross, it is easier to consider only one pair of alleles, hence, since *b* seems to be less common than *e*, let us ignore it, and consider black dogs as *EE* or *Ee*, and red ones as *ee*. The test-cross is to mate black × red. The reds can be only *ee*, but the blacks can be either *EE* or *Ee*. In genetic parlance, this test-cross is a simple backcross to the homozygous recessive type. If a black sire being thus tested is *EE*, there should be only black progeny, even though all of them will have received the gene *e* from their dam.

On the other hand, if the sire under test happens to be *Ee*, half his gametes will carry *E*, and half *e*; he should have both black and red pups in approximately equal numbers. We are not concerned here to know how closely those numbers fit the expected ratio of 1 : 1. The all-important answer to the test depends on the question, "How many black pups must a sire have to assure us that the next one will not be red; i.e., that the sire is really *EE* and not *Ee*?" A single red one is enough to prove that he is *Ee*, but will five black pups guarantee that he is *EE*?

The chance of the first (or any) pup from such a test-cross being black if the sire is *Ee* is 1 in 2. It is the same for No. 2, for No. 3, and for them all. The probability of getting (by chance alone) two black ones in succession is the product of the separate probabilities for each; i.e., $1/2 \times 1/2$, or 1 chance in 4. Three blacks can be expected $(1/2)^3$ times, or about once in 8 litters of three. Even five blacks could be expected in $(1/2)^5$ times or about once in 32 litters of five.

How many black pups are necessary from this test-cross to assure the breeder that the sire is *EE*? The answer to that question depends on another one: "With what odds will the breeder be satisfied?" He can take his choice from those given in Table 6-1. If the breeder thinks that odds of 31 : 1 are good enough, all he needs is five black pups. If he should prefer to make assurance doubly sure,[1] he will need more. Even when the chance of something happening is only 1 in 32, that event can happen, as any mother of five daughters or five sons can testify.

Test-crosses When the Homozygote Is Not Viable

It often happens that some hereditary defect that the breeder wants to eliminate from the kennel is lethal to the homozygote. The lethal action may occur during gestation, soon after birth, or at some later stage. The bird-tongue puppies (see Chapter 1) died a day or two after they were born. There is a hereditary ataxia in dogs that is usually lethal before three months of age, and gray Collies seldom live a year. Each of these abnormalities is caused by an autosomal recessive gene in the homozygous state. Only heterozygotes survive.

---

[1]If this expression perplexes you (as it did one reviewer), see Shakespeare's *Macbeth*, Act IV, Scene 1.

TABLE 6-1 Probabilities in the Test-cross of Black ♂ × Red ♀ That the Sire Is *EE*, and Not *Ee*

Offspring All Black, Number	Probability That ♂ Is Not *Ee*	
	Expressed as Numeral	Expressed as Chance
1	0.50	1 in 2
2	0.25	1 in 4
3	0.125	1 in 8
4	0.0625	1 in 16
5	0.0313	1 in 32
6	0.0156	1 in 64
7	0.0078	1 in 128
8	0.0039	1 in 256
9	0.0020	1 in 512
10	0.0010	1 in 1024

With such conditions, test-crosses have to be made by mating the suspect (of genotype unknown) to a known heterozygote. Any sire or dam that has produced affected offspring is a known carrier unless the defect is sex-linked (in which case, as we shall see shortly, the sire can usually be exonerated). With such a test-cross, the question to be answered is the same as in the previous example: "How many pups are needed to tell whether or not the suspect carries the unwanted recessive gene?" The answer, however, is not the same. In the other test, the ratio expected in the progeny (if the sire were heterozygous) was 1 : 1. In this one, the ratio expected if the suspect be heterozygous is 3 normal : 1 defective. In other words, the chance of a homozygous recessive pup showing up is only 1 in 4, whereas in the other test it was 1 in 2. Obviously, more pups will be needed. Again, it is up to the breeder to decide what degree of risk he will take; i.e., what odds are acceptable to him, and hence how many normal offspring will reduce to his minimum the risk that the suspect does not carry the unwanted gene. The figures in Table 6-2 will help him decide.

Acceptable Risks

It is clear that, in both these test-crosses, each additional pup that is black (in Table 6-1) or normal (in Table 6-2) reduces the risk that the sire under test might be carrying the unwanted gene. Biologists tend to think that a probability of 0.05 (i.e., 5 chances in 100, or 1 chance in 20) is fairly significant, but they feel much better about their conclusions when the probability that they are wrong is reduced to 0.01 or 1 in 100. To me, it would seem that, as shown by Table 6-2, one gets fairly certain with twelve normal offspring, but, because each normal pup after that greatly reduces the risk of a carrier being undetected, three or four more would not hurt at all.

TABLE 6-2 Probabilities in the Test-cross of Suspect ♂ × Known Carrier ♀ That the Suspect Does Not Carry the Autosomal Recessive Gene Under Test

Offspring All Normal, Number	Probability That Suspect Is Not a Carrier	
	Expressed as Numeral	Expressed as Chance
1	0.75	3 in 4
2	0.56	1 in 1.77
3	0.422	1 in 2.37
4	0.316	1 in 3.10
5	0.237	1 in 4.22
6	0.178	1 in 5.55
7	0.1335	1 in 7.7
8	0.100	1 in 10
9	.075	1 in 13.3
10	.056	1 in 17.7
11	.042	1 in 23.7
12	.032	1 in 31.5
13	.024	1 in 42
14	.018	1 in 56
15	.013	1 in 74
16	.01	1 in 100

All of such testing must depend in part on the average size of litter to be expected, and also on how long the pups from the test have to be kept before they can be classified as normal or not. Whereas an Irish Setter might yield a conclusive test in one big litter, it could take three litters to provide the same information in Brussels Griffons or in Pekingese. Similarly, if the unwanted trait can be identified at birth, it is easier to get two litters as a test than when all the pups have to be raised to three months of age before their phenotypes are definitely known.

We should remember that, if the suspect in either of these two kinds of test-crosses is heterozygous for the unwanted gene, a single pup of the recessive type is enough to reveal that fact.

Other Aspects of Testing

It is desirable to point out that carriers of unwanted genes can be useful, and should not be eliminated as soon as their genotypes have been identified. Without them, tests could not be made to find out what other dogs in the kennel are heterozygous for the normal allele of the unwanted gene. To be sure, such animals are identified whenever the unwanted type appears in their progeny, but, once any recessive abnormality is known to be in the stock, tests to find which dogs are carriers and which are not will accelerate elimination of the defect.

In the two preceding examples used, it was assumed that the suspects under test were both males. There is no reason why suspect females should not be similarly tested, but, since one male can sire several litters while the bitch is producing one, it seems most important to ensure that the sire is genetically sound. This is particularly important for males kept as stud dogs.

Tests for Sex-linked Genes

Contrary to what we have just read, when genes on the X chromosome are involved, breeding tests of males are not likely ever to be needed, and tests of females are essential. Any sex-linked defect, even if recessive, must show up in males, because their Y chromosome does not carry the normal alleles. An exception in which a test of males might be made would be a case in which some sex-linked defect is not fully expressed, so that, in order to be certain, another generation might be desirable. That generation would have to come from affected females or carriers.

Tests for the sex-linked genes causing hemophilia A and hemophilia B are simple. One need only mate the suspect female to any normal male and raise about six sons to the age at which hemophilia becomes recognizable. Because females that carry a gene for hemophilia A are heterozygous for that gene, the chance of any one pup getting it is ½, and the chance of any pup being a male is also ½ (approximately). The probability is thus ¼ that any pup will be both a male and hemophilic. Rather than to consider both of these probabilities combined, it is simpler to count only the males, and to remember that, by chance, about half of them should be hemophilic if their dam is a carrier. That probability (½, or 0.5) is identical with the one considered in Table 6-1. There it was applied to the passing of a recessive gene to the progeny. It applies equally well to the passing of a sex-linked gene in the dam to sons only. We see that, if the female suspected of carrying hemophilia A has six sons, none hemophilic, the chance that she might still be a carrier is only 1 in 64. The odds are thus 63 to 1 against it, which would satisfy many people, but some of us who have seen recessives crop up after none were expected would prefer to get a few more progeny just to make certain.

With sex-linked genes that are not lethal (e.g., flat front feet) a simple backcross of suspect ♀ × affected ♂ will reveal whether or not the female carries the causative recessive gene. Similar tests can be made for hemophilia B, which allows at least some of the affected males to survive to breeding age. In these two cases, pups of both sexes could show the abnormality, and the verdict is not left to the sons only, as it is with hemophilia A.

Binomials and Bitches

Many a dog breeder, contemplating prospective sales for the litter to arrive next week, will hope that there are more males than bitches. Nature seems to support such rank discrimination, for, as we saw in Table 4-1, the sex ratio at birth usually shows that males slightly outnumber females, sometimes even as much as 124 ♂♂ : 100 ♀♀. Such figures, however, apply to large numbers—not to any one litter. The concern of the breeder is more about one litter at a time, and less for the hundreds of litters needed to give some idea of the sex ratio for the species or breed.

To simplify determinations of expectations in any one litter, let us consider the sex ratio as 1 : 1, but let us not forget that we err slightly in doing so. The determination of sex is (in one sense) like a backcross to a recessive, in that the sire produces ♂-determining and ♀-determining gametes in approximately equal numbers, while all the ova are equipotential (so far as determination of sex is concerned). The probability that any one pup will be a male is again 50 : 50, ½, or 0.5, and, as we learned earlier in this chapter, the probability of getting two, three, or five of them without any females is $(1/2)^2$, $(1/2)^3$, or $(1/2)^5$. With a little simple multiplication, we find that the chance of getting only males in a litter of five is 1 in 32. The chance of getting only females is the same, so the probability of having all pups the same sex in a litter of five is 2 in 32, or 1 in 16.

Of course, such success (or disaster) may never be encountered by the breeder who raises only one litter a year. Combinations of the two sexes in one litter are the rule. There is a simple way in which the frequency of such combinations can be estimated. We can find the expectations for litters of any size from the terms of the expansion of the bionomial $(a + b)^n$

where a = probability for a ♂ = ½
b = probability for a ♀ = ½
n = number of pups in the litter

To illustrate results for the reader who has forgotten his algebra (and has no children in high school to help him with his homework), the terms in an expansion of $(a + b)^4$; that is, a litter of four, are

$$(a + b)^4 = a^4 + 4\,a^3b + 6\,a^2b^2 + 4\,ab^3 + b^4$$

To interpret all this in terms of puppies, it should be noted first that the coefficients of the five terms add up to 16. In 16 litters of four, we should expect that, by chance alone, one would be all males and another all females. In about 6 of the litters, there should be 2 ♂♂ : 2 ♀♀. About 4 litters should contain 3 ♂♂ : 1 ♀, and another 4 might be expected to have 1 ♂ : 3 ♀♀.

This is simple enough with litters of four, or even six, but, recalling that Setters often find it cheaper by the dozen, and that expansion of

$(a + b)^{12}$ takes a lot of calculation, most breeders would like to know any shortcuts that could save time and paper. Fortunately, there is one. The coefficients for any number of terms (or any size of litter) can be easily taken from Pascal's triangle, so named after the eminent French scientist who discovered it (Table 6-3).

In that triangle, each coefficient is the sum of the two nearest ones above it. The bottom line tells us that in litters of seven we might expect all males, or all females, only once in 128 litters, and that the most frequent combination is likely to be either 4 ♂♂ : 3 ♀♀, or 3 ♂♂ : 4 ♀♀, each of which should occur by chance in 35 of the 128 litters. Anyone wanting the corresponding expectations in a litter of twelve (and having enough patience and paper to find them) can easily expand the triangle to show the probabilities of getting any of the thirteen possible combinations of ♂♂ : ♀♀ in 4,096 litters of twelve.

Binomials for 3 : 1 Ratios

All of the preceding section dealt only with the expansion of the binomial $(a + b)^n$ in which the values for a and b were both ½. Dog breeders might also be curious to know what combinations to expect in litters when some simple recessive defect is segregating in a ratio of 3 dominant : 1 recessive. In this case, the exponential values for a and b are ¾ and ¼, and the binomial can be expanded as before, but a little more arithmetic is required.

Because there are still some breeders who, having read about 3 : 1 ratios, think that when three dominants have been born the next must be a recessive, it seems desirable to work out just what the expectations are. With a litter of four, the most frequent combination in the expansion of $(a + b)^4$ is (as we saw earlier) $6a^2b^2$, but in this case a is three times as great as b, and the term works out to

$$6\,(3/4)^2\,(1/4)^2 = 54/256, \text{ or } 0.2109$$

We should expect a combination of two dominant and two recessive about 21 times in 100 litters of four. Table 6-4 gives expectations for litters of eight.

TABLE 6-3 Pascal's Triangle; Coefficients of the Terms in the Expansion of the Binomial $(1/2 + 1/2)^n$

Size of Litter, n	Coefficients	Sum
	1 1	
2	1 2 1	4
3	1 3 3 1	8
4	1 4 6 4 1	16
5	1 5 10 10 5 1	32
6	1 6 15 20 15 6 1	64
7	1 7 21 35 35 21 7 1	128
etc.	etc.	

TABLE 6-4 The Probable Frequencies (Percent) of Different Combinations to Be Expected in Litters of Eight Puppies

Combination	For Dominant : Recessive 3 : 1	For ♂♂ : ♀♀ or 1 : 1
8 : 0	10.01	0.39
7 : 1	26.70	3.13
6 : 2	31.15	10.94
5 : 3	20.76	21.87
4 : 4	8.65	27.34
3 : 5	2.31	21.87
2 : 6	0.38	10.94
1 : 7	0.04	3.13
0 : 8	—	0.39
	100.00	100.00

Source: Warwick (1932).

It should be noted that the two probabilities in the binomial that is to be expanded must together equal one: ½ + ½, ¾ + ¼, and so on. Similarly, the sum of all the separate probabilities must add up to 100 percent, as in Table 6-4.

That Apparent Excess of Recessives

When the incidence of some hereditary recessive abnormality is recorded in several litters, it commonly happens that the total number of affected pups is much greater than the 25 percent expected. The ratio of normal : abnormal can even resemble 2 : 1, almost as well as the 3 : 1 expected. This is more likely to be evident when the average size of litter is only four or five, or fewer, than with large litters.

This usually happens because the breeder or the investigator analyzes only the litters that include at least one of the abnormal pups. Parents of all such litters are thus proven to be carriers of the recessive gene concerned. There could have been other litters from parents both carriers in which, by chance alone, no recessive homozygotes occurred. When the expectation is 3 : 1, litters containing none of the recessive type could be expected (by chance alone) in about 24 percent of litters of five and even 10 percent (Table 6-4) of litters of eight.

The numbers actually observed should be corrected, therefore, to find the numbers to be expected from *all* matings of carrier × carrier, including not only those which yielded recessives but also those that did not. These last, obviously, are more likely to be small litters than big ones. For that reason, it is necessary to determine separately the numbers to be expected in litters of different sizes. There are various formidable formulae for doing so, but the mind not mathematically bent can conveniently by-pass them all and use the figures worked out long ago by Bernstein (1929) for families up to 10 in man, and more recently extended by Federer (personal communication, 1975), so that they now cover litters of pups up to 12.

Use of these numbers to be expected in litters of different sizes is illustrated by the example in Table 6-5, which shows the record for 12 litters that included one or more pups having the bird-tongue lethal defect considered at length in Chapter 1. Altogether, the ratio of normal : bird-tongue was 54 : 22, which did not differ significantly from the expectation (3 : 1) of 57 : 19, but, as Table 6-5 shows, the fit of observed numbers to those expected is even better when the data are analyzed as they should be. For the convenience of other investigators, all the corrected numbers are given for litters of 2 to 12, although not all were used in this case nor are they likely to be needed in every study.

One might question the need for any such detailed analysis in the case of the bird-tongue pups because the number observed was only 3 more than that expected on a simple 3 : 1 basis. Were the corrected expectations really necessary to prove the point? Not in this case, but they could be in others, particularly when the average litter numbers fewer than the 6.3 pups per litter in Table 6-5. Subtraction of the uncorrected expectations from the corrected ones in that table shows that the correction needed is 0.643 in litters of 2, but diminishes steadily until it is only 0.098 for litters of 12.

TABLE 6-5 Number of Bird-tongue Pups Expected and Observed in 12 Litters

Size of Litter	Number of Such Litters	Affected, Expected			Affected, Observed
		Per Litter		In All Such Litters	
		Uncorrected	Corrected		
1	none	0.25	—	—	—
2	1	0.50	1.143	1.143	1
3	none	0.75	1.297	—	—
4	none	1.00	1.463	—	—
5	2	1.25	1.640	3.280	3
6	1	1.50	1.825	1.825	2
7	7	1.75	2.020	14.140	13
8	none	2.00	2.222	—	—
9	1	2.25	2.433	2.433	3
10	none	2.50	2.515	—	—
11	none	2.75	2.871	—	—
12	none	3.00	3.098	—	—
Totals	12			22.821	22

Source: Data for sizes 2–10, Bernstein (1929); sizes 11–12, Federer (personal communication).

The point is illustrated by a friend's data for five litters, each of which included one or more pups that developed a nervous disorder believed to be a simple recessive trait. Altogether, there was a ratio of 13 normal : 10 abnormal. This differed so greatly from the (uncorrected) expectation of 17.25 : 5.75 that on statistical analysis (that ever-helpful chi-square test) the discrepancy was found great enough to disprove the original hypothesis. Subsequently, when the data were analyzed by the method of Table 6-5, the corrected expectations were found to be 15.14 : 7.86 (in whole pups, 15 : 8), which were not significantly different from the 13 : 10 observed. The original hypothesis was sustained.

It will be noted that in this example, the average size of the five litters was only 4.6. There could easily have been litters of 3, 4, or 5 in the same kennel, from parents both carriers, which, by chance alone, contained none of the abnormal pups.

Part II

Genetic Variations in Dogs

With Special Reference to Abnormalities and Defects

Foreword

Chapters 7 to 12 are devoted to a survey of what is known about genetic variations in the anatomy and physiology of the dog. They were almost entitled "hereditary defects and anomalies," but the realization that some readers might not like to see under that label distinguishing features of their short-legged, short-nosed, or hairless breeds led to adoption of the comparatively innocuous title given here.

The author approaches this review with some trepidation, and warns the reader in advance not to hope to find in it the answers to all questions. There is no dearth of writings on dog abnormalities that are said to be hereditary. Some of them were listed on rather skimpy evidence, and many of them without the numbers and test matings desirable to prove whether or not the abnormality is inherited, and, if so, how. Naturally, most of the literature on defects and abnormalities in dogs is in the veterinary journals. Many of the reports describe in meticulous detail the pathology associated with the particular disease or defect under discussion, perhaps also the tests made for any possible infection, or other items. So far as genetic aspects are concerned, too many of them conclude with the belief that the condition is hereditary, probably recessive or "familial," but fail to give adequate evidence to justify that conclusion, much less to prove the mode of inheritance.

To anyone interested in canine genetics, such conclusions are a clear signal that the condition described needs further study, perhaps additional litters and test-matings. Pending real evidence, any geneticist trying to help dog breeders eliminate hereditary defects from their kennels can say only (with respect to many abnormalities) that the condition is said to be hereditary and that selection against it should be done. Actually, the number of hereditary defects in dogs proven thus far to be caused by single genes (when homozygous) is comparatively small.

Recognizing that the somewhat pessimistic views just expressed might indicate only a warped disposition in this author rather than the true state of affairs, the views of some other writers on hereditary defects in dogs were explored.

One of them (in Great Britain) listed 90 genetic anomalies, each said by someone, somewhere, to be hereditary, but gave a definite mode of inheritance for only 19 percent of them. For many of the remainder, the author (wisely) hedged, using terms such as "is said to be," "appears to display," "familial," "could be," "speculated to be," and "has been proposed" (Robinson, 1972).

Another, an American veterinarian who has contributed much to our knowledge of canine genetics, listed 104 genetic disorders of dogs for his class, and put a question mark against those for which the mode of inheritance was unknown or unproven. The proportion qualifying for that mark was 69 percent (Patterson, 1977).

A third similar list comes from Germany, and is particularly pertinent because Wegner (1975), who compiled it, made a more complete survey of European veterinary literature than one is likely to find on this side of the Atlantic. He kept his list of "congenital or juvenile anomalies" down to 50, and for 86 percent of these the cause was given as *ungeklärt,* i.e., not clarified.

With this support from three different countries, we shall try to cleave to that which is proven, and to tread only lightly on ground that is uncertain. All is not lost. During the past decade, there have appeared some excellent reports on genetic abnormalities in dogs, and there is every reason to hope that, with such examples before us, some of the dark corners of canine genetics will be elucidated in the years ahead.

7 Skeleton and Joints

The Skeleton

Size and Conformation

Great differences among breeds in size, length of leg, and shape of the head are conspicuous at any good dog show. Because the major variations are constant within breeds and constitute essential characteristics of breeds, they are clearly determined by heredity. The size and conformation preferred in any breed is maintained by constant selection to preserve breed standards and to eliminate deviations from them. That selection really seeks to maintain the desired size and type of the skeleton.

The evidence from crosses between breeds differing in size or conformation shows that variations in the skeleton as a whole are polygenic. The F_1 generation is (on average) intermediate between the parent breeds. Backcrosses are again intermediate, and in the F_2 generation there is a wide range in type.

With respect to variations in single bones, or single parts of the skeleton, the situation is different. There is ample evidence that these can be affected not only by polygenes with small, cumulative effects, but also by single mutations with great ones. The former have been utilized in selection by breeders to establish breed characteristics, such as the long nose of modern Collies; the short-faced, brachycephalic heads of Bulldogs; and the short legs of Welsh Corgis and Dachshunds.

On the other hand, mutations in single genes can induce undesirable variations in the skeleton. Of those known thus far, very few segregate in clear Mendelian ratios. Most of them show up irregularly, but do appear often enough in related animals to indicate that the tendency to produce them is hereditary in the stock. Such abnormalities are commonly said by some practitioners to be "familial," and by some geneticists to be "threshold characters." The music is the same for both—the condition is hereditary, but whether it will be manifested or suppressed in any one individual depends on the interaction of all the genes that make up the complex genotype of that individual. Some typical examples follow.

Short Spine

Short spine is a remarkable example of a mutation that causes major abnormalities in the skeleton. These are confined to the axial skeleton, which is shortened from neck to tail. The skull and the limbs are not affected. The dogs are hump-backed because of the close crowding of the vertebrae, and the spine is crooked. The tail is very short and "screw," but the neck is so short that it appears to be missing (Figure 7-1). The short-spined puppies can be recognized at birth by their very short tails. Mature dogs have difficulty in rising after they lie down flat and find it easier to relax in a sitting position with extended forelegs. As baboons adopt that same position, the short-spined dogs became known as "baboon" dogs.

(a)

(b)

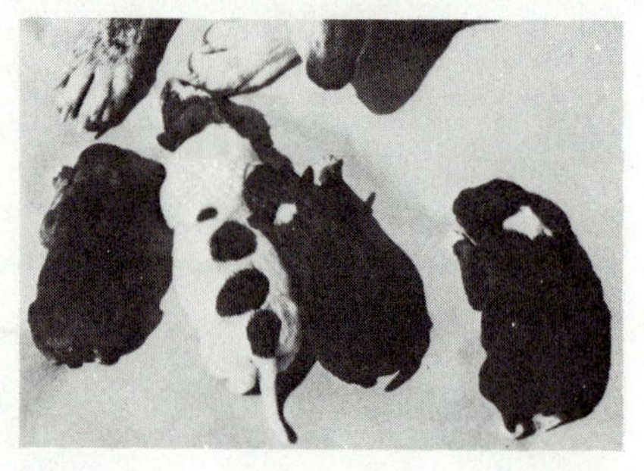

(c)

FIGURE 7-1
Short-spined "baboon" dogs in South Africa. (a) In their usual sitting position, which earned their nick-name. (b) A pregnant female showing prominent humped back. (c) Her litter by a heterozygous male showing differences between the three short-spined puppies and their normal litter mate. [Courtesy of Professor H. P. A. de Boom, Onderstepoort, Transvaal.]

FIGURE 7-2
A short-spined dog in the seventeenth century. Painted in 1690 by the Swedish artist, D. Kl. Ehrenstrahl, and described as being from a cross of fox and dog. [Photo from National Museum, Stockholm.]

This abnormality occurred in Greyhounds in Natal, and was brought to my attention by Dr. H. P. A. de Boom of the Veterinary Laboratory at Onderstepoort, in the Transvaal. In his brief report of the condition (published in 1965), he states that it turned out to be a recessive mutation.

Affected females reproduce normally but none of de Boom's short-spined males survived longer than six months. They seemed unusually susceptible to infections of their hindquarters, which resulted from their sitting upright on the ground, with forelegs extended, when at rest (Figure 7-1). It is interesting to reflect that genetic differences in ability to resist such infections may have been the instruments of natural selection by which the baboons (the real ones—not the dogs) developed over the centuries the ischial callosities so conspicuous in any baboon walking away from the viewer.

Studies of an apparently identical mutation in Japanese dogs were reported by Burns and Fraser (1966). There is also evidence that there were short-spined dogs elsewhere during the last half of the seventeenth century. Hansen (1968) has reproduced paintings of two of these, made for Charles XI of Sweden by a Swedish artist (Ehrenstrahl) who specialized in painting dogs. One of these was labelled as "a monster of wolf and dog born in Finland." Another, painted in 1690, and then supposed to be a cross of fox × dog is shown in Figure 7-2.

The apparent absence of any neck and tail, and the humped back, are remarkably similar to the conditions in de Boom's Greyhounds.

With these reports of the short-spine mutation in three widely separated parts of the world, it is clear, not only that the mutations have occurred independently, but also that somewhere, sometime, the same thing can be expected to happen again. Recessive mutations causing a short-spine condition similar to that in the dog are known in cattle and turkeys (in which they are lethal at birth or before), and in pigs, a few of which survive to maturity.

Thoracic Ectromelia

Thoracic ectromelia is an extremely abnormal condition studied in France by Ladrat et al. (1969). Both forelegs were completely lacking in the 18 dogs in which the defect was seen, but the hind limbs were normal. In Yama, the female from which all the affected pups were descended, there was a remnant of the humerus adjoining the scapula on both sides of the body. She and 8 of her ectromelic descendents are pictured in the original report.

Yama lived to five years of age and learned to stand upright, an art that seemed to raise her social status among other dogs. When provoked or teased by them, all she had to do to scare them off was to rear up on her hind legs. Among her 17 ectromelic descendants (11♂♂, 6♀♀), two were stillborn, 14 died within a week (chiefly from being unable to suckle), and only one lived to fifteen months.

The abnormality did not prevent Yama from having four litters. When bred, she was supported by the chest. Because other genetic data concerning this rare abnormality are not likely to be available, the record is given in detail in Table 7-1.

It was concluded that the ectromelia in this case was caused by a single recessive gene in the homozygous state. The authors list earlier

TABLE 7-1 Ectromelia in Descendants of Yama

Generation	Sire	Dam	Progeny	
			Normal	Ectromelic
F_1	Fox Terrier	Yama	4	0
Backcross	♂ A[a]	Yama	5	7
F_2	♂ A[a]	his sister *a*	7	1
	♂ A[a]	his sister *b*	2	8
	♂ B[b]	*a*	4	1
		Totals in F_2	13	10

[a] ♂ A was a son of the Fox Terrier × Yama.
[b] ♂ B was a son of ♂ A × Yama.

Source: Ladrat et al. (1969).

reports of similar cases, and the occurrence of hereditary "amputations" of all four legs is known in other species.

Malformed Jaws

Malformed jaws, with resultant malocclusion of the teeth, are not uncommon. Variations in the degree of shortening of the lower jaw range from extreme to barely perceptible. With one exception, these abnormalities have refused to conform to the behavior expected of a simple Mendelian character.

Stockard (1941) made several crosses between breeds that differed in shape of the head and in length of the jaws. The extreme variations found in his F_2 generations showed that they were polygenic. In Whitney's Cocker Spaniels, normal parents produced every year some puppies that developed undershot lower jaws (too long)[1] but their frequency was too low to fit expectations for any simple recessive trait (Whitney, 1972). So was that of the "pig-jaw" (lower too short) studied in Cockers by Phillips (1945), but, from his experimental mating of parents both affected, the four pups that lived all developed the "pig-jaw."

In contrast to such evidence, the short mandible studied by Grüneberg and Lea (1940) in Dachshunds did segregate nicely. In fourteen litters yielding one or more affected dogs, the ratio was 54 normal : 20 affected. In this case, the upper jaw was described as "overshot," causing malocclusion of the incisor and canine teeth. From the evidence, it seems clear that (in one breed, at least) there can be a hereditary shortening of the lower jaw that is a simple recessive trait.

Short Legs

Short legs have been adopted as a distinctive character of Pekingese, some Terriers, Basset Hounds, Dachshunds, and some other breeds. They result from achondroplastic (chondrodystrophic) changes in the long bones of the legs. In some breeds, these are associated with changes in the skull, but in others the legs are shortened without any corresponding modifications in the head. Crosses indicate that, in general, short legs are incompletely dominant to long ones, but evidence is lacking to confirm Stockard's opinion that a single gene is responsible. Whitney's crosses of short-legged × long-legged breeds yielded progeny that were intermediate. Backcrosses of an F_1 male to both types

---

[1]Being uncertain whether that part of the biting equipment said by breeders to be "overshot" or "undershot" is the mandible (lower) or the maxilla (upper), I am relieved to find that even Whitney, after breeding dogs all his life, was unable to decide "whether the trouble lies in the mandible being too short or the upper jaw too far forward." Perhaps the safest policy for a writer to follow is that of Humpty-Dumpty *(Alice in Wonderland):* "When I use a word . . . it means just what I choose it to mean—neither more nor less."

resulted in a continuous gradation of leg length in dogs from the short-legged bitch, but more distinct differences in length of leg among those from longer-legged females.

Whitney found in Cocker Spaniels a recessive type of short legs. Normal parents (closely related) produced in four litters 22 offspring among which 6 had conspicuously short legs. There was at least 1 in each litter. When mated with another relative, the sire produced another short-legged dog, but, among his progeny (numbering over 100) from other bitches, none of the short-legged type appeared. Apparently, one kind of short legs can be a simple recessive trait.

It seems safest to conclude that length of leg is determined by an unknown number of genes, but that among them there can be some that exert more influence than others.

Tails

Tails show more variation than other parts of the skeleton, and it is not surprising to find that most of the known variants have been adopted as breed characters. Nor is it surprising that they have attracted a certain amount of genetic investigation, or that the writers on *anury* (taillessness), brachyury (short tail), and screw-tail are not in complete agreement on the numbers and kinds of genes responsible for the variation. Readers interested are referred to Burns and Fraser (1966) for a review of this subject.

They should also consult Whitney's books (1948, 1972) for details of his many crosses involving tails of different lengths. The 11 different types of tails that he recognized are shown in Figure 7-3. In addition to these, he found some abnormalities (screw, "jog," curled, and so on) that appeared to be non-genetic. Such things result from accidents during development of the embryo.

From the constancy with which different kinds of tails are maintained as breed characters, it is clear that the variation among them is genetic in origin, but speculation about the numbers and kinds of genes responsible for that variation is a pastime in which I do not care to indulge.

Cleft Palate

Cleft palate is found, along with other abnormalities, in some syndromes, but can occur without other complications. There are suggestions that it is genetic in origin, but most of such claims lack adequate evidence. Wriedt (1925) reported a strain of Bulldogs in which six litters from matings said to be of carrier × carrier yielded a ratio of 24 normal : 9 with cleft palate. The defect caused the mother's milk to run out through the noses of the nursing puppies, and most of them starved to death. In spite of the nice 3 : 1 ratio in this case, one must not conclude that cleft palate is a simple recessive character.

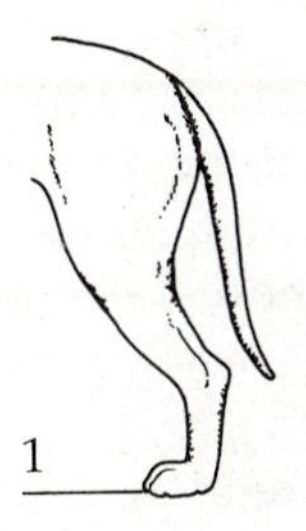

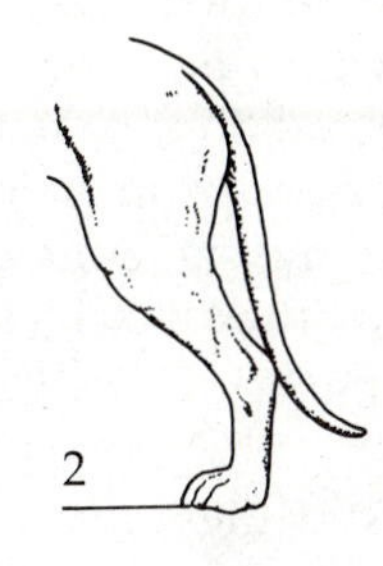

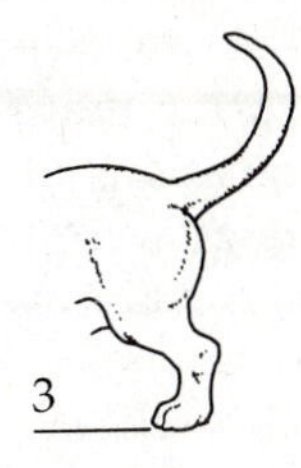

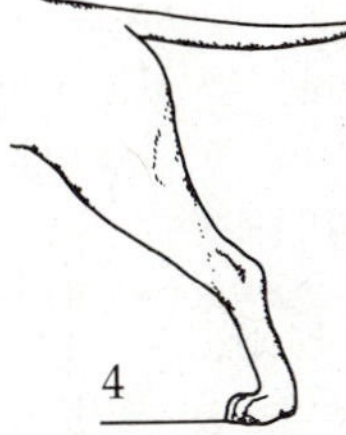

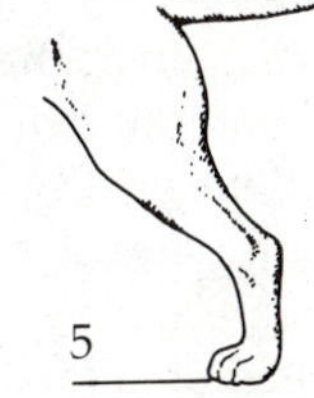

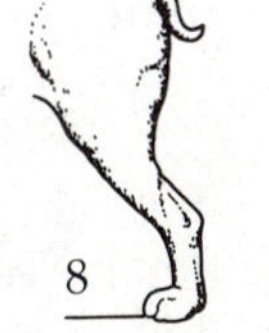

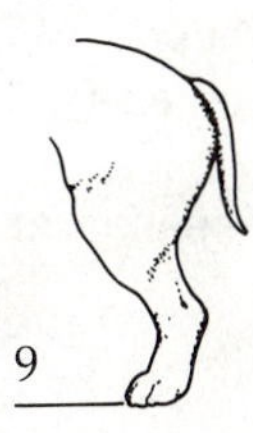

FIGURE 7-3

Genetic variations in tails. Breeds exemplifying are given by Whitney as 1, several; 2, Borzoi; 3, Beagle; 4, Pointer; 5, some hounds; 6, Cocker Spaniel (occasionally); 7, Old English Sheepdog, Schipperke; 8, Boston Terrier; 9, English Bulldog; 10, Pug; 11, Eskimo, Chow. [Reprinted from *How to Breed Dogs,* copyright © 1971, 1947, 1937 by Leon F. Whitney, D. V. M., with special permission of the publisher, Howell Book House, Inc.]

According to Weber (1959), dominance seemed more likely when a Bernese Sennenhund ♂ , which had a median nasal cleft, sired 11 offspring that had the same condition, or harelip, or cleft palate, or combinations of these, among 26 pups in four litters. Some of the affected dogs were from a female of the same breed, and 3 were from an unrelated German Shepherd, neither of which had any of those defects.

Dew-claws

It is not clear why the extra digit that is commonly found in many species should be called a *dew-claw* when it occurs in the dog. According to Winge (1950), it is known in Europe as a "wolf claw," but, since wolves normally have no more toes than the dog, there seems little justification for charging it against the dog's relatives in the wild. Dew-claws are sometimes bilateral, but often unilateral. According to Whitney, most dogs have them on the front legs. Sometimes there is only a vestige of the extra toe, and it may lie in the skin or under it. Dew-claws on the hind legs are usually clipped off at birth by fanciers except when required by breed standards. Great Pyrenees should have double ones behind and single ones in front.

Some investigators have found dew-claws to be dominant, others recessive. The conflicting evidence on possible genetic bases is reviewed by Burns and Fraser (1966).

Diseases of the Spine

Apart entirely from the extremely abnormal short spines described earlier, dogs can suffer from other, less-conspicuous afflictions of the spine. Among these are the following.

Slipped Disc

Slipped disc, often called *protrusion of intervertebral disc,* or simply *disc disease,* has been reported in several breeds, but is apparently more frequent in Dachshunds than in any other. Among 782 cases of it in Texas, Gage (1975) found that 77 percent were in Dachshunds. Poodles were high on the list. It should be clearly understood that none of Gage's useful data establish the incidence of slipped disc in Dachshunds, or in any other breed. They deal only with the cases brought to him for attention. However, unless Texas is overrun with Dachshunds that do not have slipped discs, it seems fair to conclude that the incidence is unusually high in that breed. Among breeds contributing fewer than 0.1 percent to Gage's 782 cases were the Collie, Welsh Corgi, Boxer, Labrador Retriever, and Pomeranian. Among the 548 afflicted Dachshunds, 86 percent had slipped disc in the thoraco-lumbar vertebrae, and in only 14 percent were the lesions in the cervical vertebrae. The disease is associated with aging, the mode in Gage's cases being five years.

Lest any lover of Dachshunds should protest that Gage's report is an unjustified slur against that breed, and claim that something adverse in the Texas environment must be responsible rather than the genes that make the breed, it should be reported that in far-off Australia, Johnston and Cox (1970) found, among 178 cases of "luxation of intervertebral disc," that 62 percent of them were in Dachshunds.

It is difficult to select against any genetic character that is not manifested until middle age, but it is clear that dogs with slipped disc should not be used for breeding.

Spondylosis

The terms *spondylosis deformans* and *vertebral osteophytosis* are used by Morgan et al. (1967) for a condition in which bony spurs (osteophytes) occur on the vertebrae. These range in size from single small projections to complete bony bridges from one vertebra to another. Vertebrae thus rigidly joined are said to be *ankylosed.* The smaller bony spurs represent early steps in the process of ankylosis, which advances with age.

In radiographic examination of 1,451 dogs of various breeds in Sweden, Morgan et al. (1967) were able to get some idea of the incidence of spondylosis in different breeds. That was possible because the dogs they studied had not been sent in specifically to be examined for spinal abnormalities and hence could be considered a fair sample of the breeds that they represented. Some of their data are given in Table 7-2 to show how greatly the breeds differed in the incidence of spondylosis.

Another reason for showing the frequencies in breeds is to point out that the Dachshunds and Poodles, which rank highest in the incidence of slipped disc, have comparatively low frequencies of spon-

TABLE 7-2 Incidence of Spondylosis in Several Breeds

Breed	Dogs Examined, Number	With Spondylosis	
		Number	Percent
Boxer	85	46	54.3
German Shepherd	136	34	25.0
Airedale	54	13	24.1
Cocker Spaniel	110	24	21.8
Dachshund	415	26	6.3
Poodle	121	6	5.0

Source: Data of Morgan et al. (1967).

dylosis. Philosophers among us will declare it only fair that breeds specially susceptible to some diseases should be spared from others. It seems more likely, however, because these two diseases of the spine occur at about the same age (after two or three years), that the tendency to grow bony spurs on the vertebrae counteracts in some way any tendency toward degeneration of the discs between them.

In Switzerland, Mühlebach and Freudiger (1973) found, in 324 Boxers examined radiographically, that the proportion showing some degree of spondylosis was 92 percent. Their five grades of the disease ranged from osteophytes just starting to grow (Grade I) to complete ankylosis of two or more vertebrae (Figure 7-4). At 9 to 12 months of age, 23 percent of the dogs showed no spondylosis, and the rest had lesions of Grade I, but, with a single exception, all those over four years of age had severe stages of the disease. The region of the spine most commonly affected was that of the last two thoracic vertebrae and the contiguous ones (both fore and aft)[2] where there is a great dorsal-convex bend in the spine (Figure 7-4).

The Swiss investigators attributed the high incidence of spondylotic lesions at that site to the typical shape of the body; that is, to the curvature of the spine. From the fact that "families" differed in the frequency and severity of spondylosis, it was concluded that genetic influences have a role in the causation of that disease. Although their families were comparatively small, the very high frequency of the disease in Boxers is enough to confirm that assumption.

Considering together the two diseases of the spine just reviewed, one is tempted to surmise that backaches in the canine world might be fewer if breeders could be persuaded to tighten up the spines of the Dachshunds and to loosen up those of the Boxers.

Other Diseases, Possibly Hereditary

In this section are listed three abnormalities of the skeleton that might be hereditary but for which further evidence on that score is desirable.

Polyostotic Fibrous Dysplasia

The name *polyostotic fibrous dysplasia* was given to a condition found by Carrig and Seawright (1969) in three of six Doberman Pinschers that survived after three others had died at birth or soon after. It was recognized clinically by swelling and lameness in the forelegs after four months of age. Radiographic examination revealed osteolysis and formation of cysts in the distal half or third of the ulna. Because of the three cases in one litter, and the fact that the sire's sire was the dam's grandsire, the condition was described as familial. The study was made

[2]Anatomists, for *fore* and *aft,* please read *cephalad* and *caudad.*

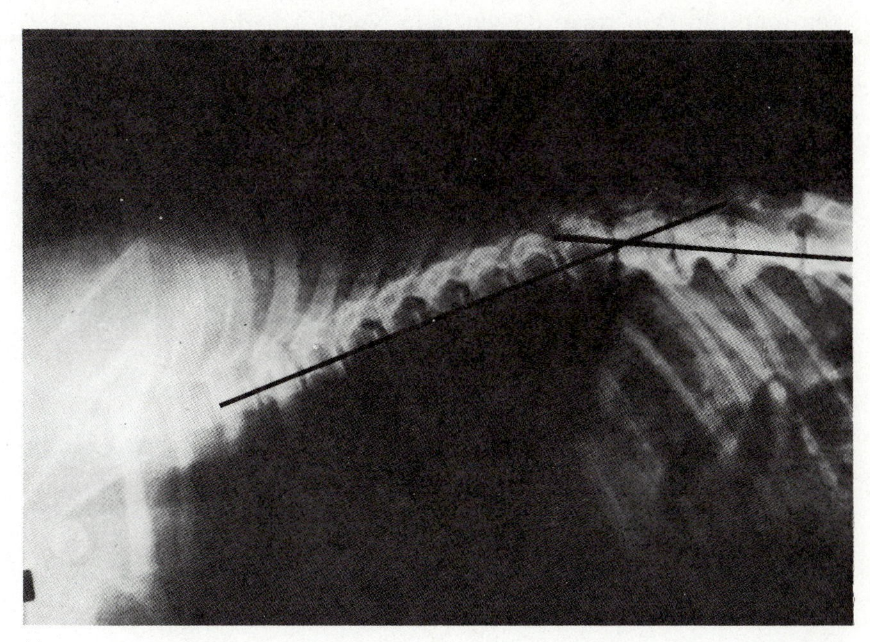

(a)

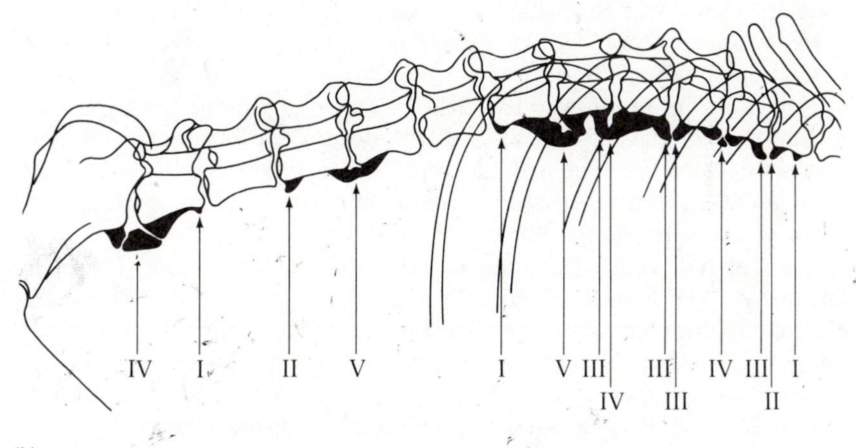

(b)

FIGURE 7-4

Spondylosis in the Boxer. (a) radiograph showing dorsal-convex bend in the spine, the most common site of spondylotic lesions. (b) Osteophytes (in black) of Grades I to V. [From Mühlebach and Freudiger, 1973.]

in Queensland, but, according to the report, the same condition has been found in Doberman Pinschers in North America.

Multiple Epiphyseal Dysplasia

The name *multiple epiphyseal dysplasia* was given by Rasmussen (1972) to a condition found in a litter of eight Beagles, four of which had malfunctioning hind legs and died early. The other four were normal. The afflicted puppies were described as having punctate calcification or stippled epiphyses, bilaterally, in the femur, humerus, metacarpals, metatarsals, carpals, and tarsals. It would be interesting to know whether or not the sire and dam were related, but on that point the record is silent.

Brachydactyly

In a strain of short-toed (brachydactylous) dogs, a mutation (said to be recessive) reduced the size and function of the outer toes on the front feet, and sometimes on the hind feet as well. Genetic data were not given (Green, 1957).

Joints

There are a few genetic abnormalities that have to be classified as defects in joints, but for which the exact causes have yet to be determined. One example is hip dysplasia, which has already been discussed in Chapter 5, and to which we shall return again in Chapter 14, on selection. In addition to the clear evidence of genetic predisposition to that condition, at least half a dozen other contributing causes have been suggested by half a dozen investigators. We have yet to learn what anatomical and environmental influences combine with genetic predisposition to induce hip dysplasia.

Somewhat similarly, subluxation of the carpal joint, otherwise labelled as "flat front feet" (see Chapter 4), has thus far defied the best efforts of competent anatomists and pathologists to find any associated abnormality other than the sex-linked gene which causes the luxation.

Three other hereditary defects of joints can be listed: luxation of the patella, elbow dysplasia, and subluxation of the elbows.

Luxation of the Patella

The patella is known to most of us as the kneecap, so its luxation in dogs is an affliction of the hind legs. Normally it glides in a rather shallow groove at the lower end of the femur as the leg is bent or straightened. A ridge on each side of the groove helps to keep it in place, but sometimes, if the groove is too shallow or the ridges too low, the patella slips to one side or the other, and the joint is thus dislocated. If that happens on one leg only, the dog walks (or runs) on three legs and holds up the other. When the joint is dislocated in both legs, movement is difficult, but is possible in a crouching position, and some dogs manage to go short distances on their front legs only. The dislocated patella can be restored to its proper place, but it does not stay there long before dislocation recurs. The defect usually does not become evident until the dog is four to six months old.

Luxation of the patella is more prevalent in miniature and toy breeds than in larger ones, but Hodgman (1963) reported its occurrence in 43 breeds in Britain. Among his 616 cases, over a third were in Miniature Poodles and about 7 percent in larger Poodles. In the records from ten veterinary colleges in North America analyzed by Priester (1972), there were 34 breeds listed with the defect. Among these were three St. Bernards. Because both of these reports are based only on the cases brought to veterinary clinics, neither can be used to determine the incidence in any breed. There is some significance, however, in the fact that, among Priester's total of 542 cases, no fewer than 243 were in Miniature or Toy Poodles. Similarly, in 169 cases of luxation of the patella in Australia, Johnston and Cox (1970) found 51 in Poodles.

From a pedigree showing 17 cases of this defect in three generations of King Charles Spaniels, Loeffler and Meyer (1961) concluded that predisposition to it is inherited and polygenic. Accordingly, breeders should select against it, as against any other polygenic trait, by not breeding from parents or litter-mates of affected dogs, or from any in which the defect may appear to have been overcome by surgical intervention.

Elbow Dysplasia

Elbow dysplasia is not quite the counterpart in the forelimb of hip dysplasia in the hind one, but both have in common the fact that they cause lameness and accompanying awkward gait. Hip dysplasia occurs at the juncture of the limb with the pelvis; elbow dysplasia occurs at the lower joint, where the humerus meets the radius and ulna.

The two conditions also differ in that dysplasia in the elbow joint is easily recognized by lameness of a front leg at an early age, whereas hip dysplasia can be recognized clinically only in severe cases, and often can be detected only by radiographs after a year of age. Furthermore, the trouble in the elbow can usually be easily corrected by surgery, but that in the hip is more complicated. Hip dysplasia is common; elbow dysplasia is not.

Elbow dysplasia results when the anconeal process of the ulna fails to ossify normally and to fuse with the ulna as it should do. Lameness is apparently recognizable sometime after about four months of age.

Cawley and Archibald (1959) found the average age at recognition in 11 German Shepherds to be six months and also that about a third of them had both legs affected. Surgical removal of the triangular bit of loose bone caused the lameness to disappear in about 10 days.

Reports of this condition in the veterinary literature are somewhat confusing. In North America, elbow dysplasia seems to have been diagnosed most commonly in German Shepherds. By contrast, Hodgman's survey of 98 cases reported in the British Isles shows 56 to have occurred in Pekingese. In the records of Pobisch et al. (1972) for more than 60,000 dogs examined over 20 years from the region around Vienna, only 62 cases were diagnosed, most of them in German Shepherds. Reports in North America suggest a much higher frequency. Some of the discrepancy results from the fact that the Vienna records considered only adult dogs aged 18 months to three years, whereas in Canada and the United States the age at recognition is much younger. It is possible that some of the afflicted dogs were eliminated for lameness at earlier ages and thus escaped inclusion in the records of Pobisch et al. In any case, frequencies and the incidence in different breeds are unreliable if based only on the number of accessions in veterinary clinics, and not on larger samples of the breeds from which those accessions come.

The relatively high incidence in German Shepherds and appearance of elbow dysplasia in related dogs suggest that susceptibility to the condition is hereditary. In quest of the genetic basis, Corley et al. (1968) made experimental matings (in German Shepherds) in which both, one, or neither of the parent dogs had elbow dysplasia. The progeny (excluding those that died earlier than 90 days of age) were examined both radiographically and by "gross pathologic examination." Some that did not have an ununited anconeal process did have arthritis attributable to elbow dysplasia. The results clearly showed that susceptibility to the condition is genetic (Table 7-3). Most geneticists will hardly agree with the conclusion by Corley et al. that three dominant genes are responsible. It seems more likely, considering the absence of Mendelian ratios and the gradation in severity of the defect, that the condition is polygenic.

Selection against elbow dysplasia should follow the same general rules as for luxation of the patella and any other polygenic defect, but in this case, since surgical correction of the defect is apparently simple and effective, breeders must be especially on guard against the use of animals that have been thus "cured."

Subluxation of the Elbows

Lameness in the foreleg can also be caused by what Lau (1977) has called "premature closure of the distal physis." It results from abnormal growth of the radius or ulna, the two long bones of the foreleg.

TABLE 7-3 Elbow Dysplasia in Experimental Matings

Parents	Progeny			
	Normal	Ununited Process	Arthritic	Proportion with Elbow Dysplasia, Percent
Both normal	17	0	0	0
One dysplastic	28	6	3	24.3
Both dysplastic	32	13	11	42.8

Source: Data of Carley et al. (1968).

Affected dogs are lame, and, when walking, tend to swing out their elbows. These joints cannot be fully extended.

During a 5-year period, Lau found this condition in 20 related Skye Terriers. From four litters in which some or all puppies were affected, he concluded that the defect is inherited, probably as a simple recessive trait. His data, however, do not rule out the possibility that it is polygenic. As it does not always affect both legs equally, and varies in severity, the latter possibility cannot be excluded.

8 Coats and Colors

After variations in size and shape, the most conspicuous features of dogs are their coats and colors. Accordingly, before reviewing genetic abnormalities inside the skin, it is appropriate to say a little about what we can see from outside it.

Since its domestication, the dog has been subject to selection by breeders whose needs, objectives, and standards varied greatly in different parts of the world. As a result, it is not surprising that breeds differ greatly in their outer covering. The hair may be long, intermediate, short, or, as in the naked breeds, almost completely absent. It can be straight or curly, thin or almost woolly. In color, it can be black or red, with dilutions of either in varying degrees. It can be solid ("self") color or show patterns with various amounts of white. This book is less concerned with such variations among breeds than with genetic defects that may occur in any breed or in all. For a review of variations among breeds in skin and hair, readers are referred to the book by Burns and Fraser (1966).

Coats

Short Coats and Long Ones

In some breeds short coats are preferred, and long-coated dogs are outcasts. In others, the two styles are considered as representing within the breed two separate varieties—rough and smooth, or long-haired and short-haired. In St. Bernards both kinds are accepted as respectable, fully-accredited representatives of that famous breed.

From reports of 16 breeders, added to data from the senior author's own kennel, Crawford and Loomis (1978) obtained classifications of the coat for 1,216 St. Bernards in 221 litters. These came from four different types of matings. They showed that short coat was dominant to long coat, and that the difference is determined by a pair of autosomal alleles, *L* and *l*. It seems probable that these same genes are responsible for the two types in other breeds.

According to Crawford and Loomis, the type of coat is clearly recognizable in St. Bernard pups at 6 weeks of age by differences in the length of hair. Moreover, while short-coated pups have bluish skin, with hair gray at the base, in long-coated pups the skin is pink and the hair is buff at the base.

Lethal Hairlessness

Differing degrees of hairlessness have been found to be hereditary in many mammals, so it is not surprising that they occur also in the dog. One extreme type has even been made the distinguishing characteristic of the breeds known in various parts of the world as Persian, Chinese, Turkish, or Mexican Hairless. Actually, these dogs are not completely hairless and have varying amounts of short, fine hair on the head, tail, and feet.

Letard (1930) confirmed previous indications that hairless dogs are heterozygous for that characteristic and, when bred to normal dogs, produce about equal numbers of hairless and normally haired progeny. From seven experimental matings of naked × naked, he obtained only 25 pups. Among these, some of the naked ones were born dead, with extreme abnormalities of the buccal cavity and lacking external ears. Others were born alive, but died within a few days because of complete occlusion of the esophagus. Some were born naked, but otherwise normal and viable. Among the 25 pups, the ratio of normal (with hair) : naked (viable) : naked (nonviable) was 9 : 12 : 4. This is a good fit to the expected ratio (1 : 2 : 1) of 6.25 : 12.5 : 6.25 from segregation of a dominant gene for which both parents were heterozygous and that was lethal to the homozygote. The possibility that some of the homozygotes may have died too early to be recognized is not excluded.

Hypotrichosis, Not Lethal

A less extreme degree of hairlessness was seen in Miniature Poodles and described by Selmanowitz et al. (1970). Extensive areas devoid of hair occurred on the head, ventral trunk, dorsal pelvis, and the upper parts of the limbs. In one litter from normal parents, two male pups were thus affected, but three females were all normal. A similar (but less extreme) case of congenital hypotrichosis in a Whippet was described by Thomsett (1961). In neither of these case reports was there any genetic evidence, but the inference in both was that, because the parents were normal, the hypotrichosis was probably heritable and recessive. Perhaps so, but a dominant mutation can appear in the offspring of normal parents.

Necrosis of the Toes

A peculiar, lethal disease was studied by Sanda and Krizenecky (1965) in Short-coated Setters in Czechoslovakia. At about four months of age, the afflicted puppies began to lick their toes and (later) to bite them. Claws were lost and eventually single phalanges also. One or more limbs were affected. A peculiar characteristic of the disease was the loss of sensitivity in the distal ends of the affected limbs, which became completely insensitive to cuts, pricking, and burning with a hot iron. Because of that insensitivity, the dogs maintained a normal gait, and did not limp even when the feet were badly diseased.

Details of the course of the disease and of the pathological changes found by x-ray and histological examination are given by Sanda and Pivnick (1964) for eight cases (including only one female) examined by them. Some lived for several months but eventually died from complications following infection of the wounds in the feet.

The fact that this condition was seen only in one breed suggested that it was probably genetic in origin. Unfortunately, because none of the afflicted dogs lived to sexual maturity, experimental matings with them were not possible. Pedigrees of 26 cases all traced back to one pair of ancestors. These appeared, sometimes repeatedly, on both the dam's and sire's side of the pedigrees.

Although it is clear that there is a genetic basis for this disease and that it is recessive, it has not been shown that a single recessive gene is responsible. As in any case in which the afflicted animal does not survive to reproduce, evidence for a simple recessive trait would be most easily provided by a 3 : 1 ratio of normal : abnormal in litters that have at least one puppy showing the defect. Such evidence is not as easy to get as one might suppose, because both dog breeders and veterinarians tend to concentrate their attention (and records) on the abnormal animals and fail to record the numbers and sexes of their normal littermates.[1] Apparently this was the case with necrosis of the toes in those Czech Setters. The question has yet to be answered whether the condition is always manifested by homozygotes (that is, a simple recessive trait) or whether its penetrance is incomplete, so that the frequency in litters from parents both carriers is less than 25 percent. It is also not clear whether both sexes are equally affected. Finally, the possibility is not excluded that the necrosis may be caused by some neurotropic pathogen to which the afflicted animals are genetically susceptible.

---

[1]A failing not restricted to the veterinarians in Czechslovakia.

Dermoid Sinus in the Rhodesian Ridgeback

This is a good example of a condition that is evidently hereditary, because it is found only rarely except in the one breed. It is believed to be associated with the ridge of hairs, on the midline of the back, that is a distinguishing characteristic of the Rhodesian Ridgeback (Figure 8-1). That ridge is conspicuous because the hairs slope toward the head. According to Mann and Stratton (1966), such a ridge was also prevalent in the Hottentot dogs that were ancestors of the Ridgebacks.

For most of us, it will be simpler to think of these sinuses as cysts in the skin. They do not occur within the ridge of reversed hairs, but either anteriorly to it, on the neck, toward the head, or posteriorly, toward the tail. There may be a single cyst or several. Hofmeyer (1963) reports having seen as many as eight on the neck, and up to three in the sacral region. According to him, the condition is present at birth and can be felt as cords running from the skin to the spine. It is believed to be caused by incomplete separation of the skin from the spine during embryonic development. The sinus has a small external opening, which (according to Hofmeyer) is about 1 millimeter in diameter in adults, and surrounded by a tuft of hair. Eventually it becomes infected, develops an abscess, and has to be cut out.

Attempts to find the genetic basis for these cysts have not been successful. Hofmeyer (1963), noting that they could occur not only in pure Ridgebacks but also in crosses with other breeds, believed it to be a dominant trait with incomplete penetrance. Mann and Stratton (1966) analyzed the records for 44 litters from Ridgebacks not cystic (in Britain) and found the incidence of pups showing dermoid sinus to be only about 12 percent. In 16 litters (118 pups), there were none at all. When the data were restricted to 28 litters that either contained one affected pup or came from parents that had previously produced one, the incidence rose to 18.7 percent. It was practically the same in 43 pups from an affected dam mated to dogs that were normal but known to be carriers. Mann and Stratton wisely refrained from trying to fit these data to the Procrustean requirements of single genes and concluded that the genetic basis for dermoid sinus is complex, i.e., polygenic.

After the hereditary nature of the defect had been recognized, breeders of Rhodesian Ridgebacks in South Africa tried to eradicate the defect. Beginning about 1942, affected dogs (including those "cured" by operation) were banned from shows, and puppies found to have the condition were destroyed. Hofmeyer (writing about 20 years later) reported that these measures effectively reduced the number of dogs brought to him for operation. Mann and Stratton, however, pointed out that, although such elimination of affected animals causes a fairly rapid decrease of any hereditary defect when the original frequency is

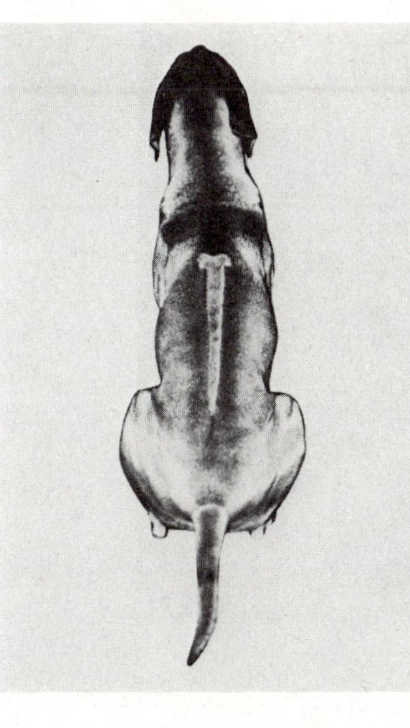

FIGURE 8-1
A Rhodesian Ridgeback, showing the distinctive ridge made by hairs that slant forward instead of backward. The lateral extensions at the top are known to breeders as *crowns*. [Reprinted from *How to Breed Dogs*, copyright © 1971, 1947, 1937 by Leon F. Whitney, D. V. M., with special permission of the publisher, Howell Book House, Inc.]

high, it does not eliminate the condition altogether, because recessive genes inducing the defect are maintained by undetected carriers. They recommended that such carriers be identified by progeny-tests and eliminated. So do I.

Ehlers-Danlos Syndrome

A peculiar disease of the dog's skin has been thoroughly investigated by Hegreberg et al. (1969, 1970a, 1970b) at Washington State University. The most obvious symptom is an extreme looseness and fragility of the skin. It can be pulled away from the underlying tissues, and is also easily torn so that the skin of affected dogs shows many broad, pliable, and thin scars. These are most numerous in parts of the body subject to injury such as the back, head, ears, and legs (Figure 8-2).

A similar condition in man was first called *cutis laxa*, but has since been given several other names (mostly polysyllabic) and is now generally known as the Ehlers-Danlos syndrome after the investigators who first studied it early in this century. When two Springer Spaniel litter-mates showing this condition were found, the Pullman, Washington, veterinarians got four unaffected females of that same breed and, with the six, made a series of matings to find the genetic basis for the disease. The evidence therefrom (summarized in Table 8-1) showed clearly that the condition was caused by a single dominant gene in the heterozygous state, and that (in this breed) penetrance (manifestation) of the trait was complete in all heterozygotes.

FIGURE 8-2
A Springer Spaniel, showing on the back and rump the scars in the skin caused by a genetic condition resembling the Ehlers-Danlos syndrome in man. [From Hegreberg et al., 1969.]

The relationships involved in seven different matings and the progeny from each are shown graphically in Figure 8-3. This figure might also be useful as a model for portrayal of results in genetic investigations of other conditions.

Study of Table 8-1 and Figure 8-3 brings out some interesting points for dog breeders to recognize:

1. When both parents were affected, the average number of pups in the two litters was only 2.5. In the other matings, even after exclusion of 7 pups that died too early to be classified, it was twice as high. Could the syndrome be lethal to homozygotes at some stage of development too early to be recognized? If so, the ratio to be expected at birth was not 3 : 1, but 2 : 1. Unfortunately, two small litters are not enough to answer that question, and the fact that the numbers observed fit a 2 : 1 ratio slightly better than one of 3 : 1 is completely insignificant in 5 pups. Further studies are necessary before the question can be answered.

2. Since both III-3 and III-5 were normal, II-1 must have been heterozygous. II-2, with 16 normal progeny in eight litters, was clearly the same. So was III-27. As the 3 affected dogs tested were all heterozygous, the possibility that homozygotes never survive is strengthened.

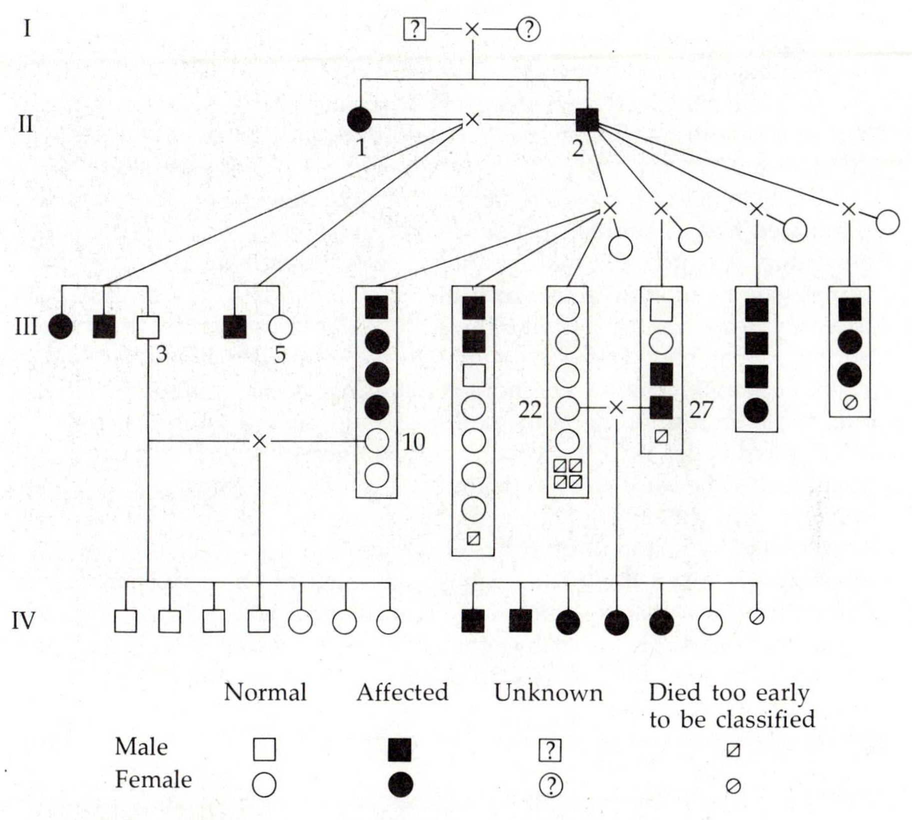

FIGURE 8-3
Inheritance in Springer Spaniels of a disease resembling the Ehlers-Danlos syndrome of man. [Redrawn, with different identifying numbers, from data and charts of Hegreberg et al., 1969.]

TABLE 8-1 Experimental Matings Showing the Genetic Basis for the Ehlers-Danlos Syndrome in Dogs

Type of Mating	Progeny				
	Litters	Pups	Affected	Normal	Expected Ratio
Affected × affected	2	5	3	2	3.75 : 1.25 *or* 3.33 : 1.67
Affected × normal	7	35[a]	20	15	17.5 : 17.5
F_1 normal × related normal	8	8	0	8	0 : 8

[a]Excluding 7 pups that died too early to be classified.

Source: Data of Hegreberg et al. (1969).

3. The normal ♂ III-3 × normal ♀ III-10 (litter-mate of four affected pups) yielded only normal offspring, eight in number. The odds against that happening by chance if either parent carried a causative gene are 255 : 1. Clearly, when a single dominant gene is involved, it is perfectly safe to breed from the normal litter-mates of pups that show the defect.

4. Obviously, simple dominant defects like this one are more easily eliminated from a kennel than simple recessive ones, or those that are polygenic. It is possible that in some breeds a dominant gene may be suppressed by modifiers, in which case some heterozygotes may not be so easily recognizable as with the Ehlers-Danlos syndrome in these Springer Spaniels.

5. Perhaps, if there were more genetic analyses as thorough as this one, the literature on heredity in dogs would be less "iffy."

6. Finally, let no breeder point the finger of shame at Springer Spaniels just because this mutation occurred in that fine breed. It can occur in any other. The Ehlers-Danlos syndrome is now known in man, minks, cats, and dogs. It has also been found in cattle in Belgium, where it was given the distinguishing sobriquet of "dermatosparaxis." Evidently it is a disease—not just of Springer Spaniels or of dogs—but of mammals. So are many others in the list of heritable defects in dogs. Fortunately, the Ehlers-Danlos syndrome seems to be one of the rarer ones on that list.

Colors

Much has been written about the inheritance of coat color in dogs. On some points, there is agreement; on many, there is not. As I have not myself contributed anything to either the facts or the controversies in this field, it seems best now to refer readers who want to know more about genes affecting colors to the writings of authors who know more about them than I do. (In my opinion, this side-stepping shows commendable reluctance to swim in murky waters, but readers who call it passing the buck may be right.) There are whole books on the subject, whole chapters in other books, and many published reports of studies with specific breeds or of crosses between them.

A good place to start is with the conclusions reached by Little (1957) from more than 40 years of studying inheritance in dogs. A summary of the genes that he considered responsible for the principal colors and patterns lists 24 genes at nine different loci (Table 8-2).

Fourteen of those genes are in series of multiple alleles at four different loci. Readers who have not read earlier chapters of this book can find out about multiple alleles in Chapter 2. An important point to remember is that in such series the alleles are usually listed in order of dominance, so that any one gene will be dominant to those below it but recessive to those above. At the five loci with paired genes, the dominant partner is, as usual, designated by a capital letter.

With the genes thus postulated, Little assigned genotypes to many breeds. Examples are as follows:

Labrador Retriever (black)	A^s	B	C	D	E	g	m	P	S	t
Cairn Terrier (light gray, brindle)	a^y	B	c^{ch}	D	e^{br}	g	m	P	S	t
Dachshund (black, tan points)	a^t	B	C	D	E	g	m	P	S	t
Weimaraner	A^s	b	C	d	E	g	m	P	S	t

In these examples only single genes are shown. Pure breeds should homozygous at each locus, but could be heterozygous in breeds allowed variations in color.

TABLE 8-2 Little's Major Genes Affecting Coat Color in Dogs

Locus	Gene	Effects of the Gene	Breeds Exemplifying
A	A^s	Self color. Distributes dark pigment over whole body	Newfoundland, Chesapeake Bay Retriever
	a^y	Restricts dark pigment; makes sable or tan	Basenji, Irish Terrier
	a^t	Induces black-and-tan, and other bicolor varieties with tan points	Dachshund, Airedale
B	B	Induces black color	Any solid black
	b	Induces liver or chocolate brown	Weimaraner, Poodle
C	C	Necessary to form any color	Blacks, deep reds, dark livers
	c^{ch}	Reduces red pigment, but with "little or no effect" in solid blacks	Schnauzer, Norwegian Elk Hound
	c^a	Complete pink-eyed albinism, very rare	None
D	D	Causes intense pigmentation	Irish Setter, Schipperke, and others
	d	Dilutes black pigment to blue	Blue varieties in Great Dane, Greyhound, and others
E	E^m	Induces "black mask" pattern	Pug, Norwegian Elk Hound
	E	Extends dark pigment throughout coat	Cocker Spaniel, black or brown
	e^{br}	Brindle, but interactions with other genes are not clear	Boston Terrier, Greyhound
	e	Restricts black to eye, leaving coat red, or with shades to yellow	Cocker Spaniel, Golden Retriever

(continued)

TABLE 8-2 Little's Major Genes Affecting Coat Color in Dogs *(Cont.)*

Locus	Gene	Effects of the Gene	Breeds Exemplifying
G	*G*	Progressive graying from black at birth	Some Kerry Blues and other breeds
	g	No such graying	Most breeds
M	*M*	Merle or dapple pattern, often with white areas	Merle Collie and others
	m	Uniform pigmentation	Most breeds
S	*S*	Solid color, no white	Any with solid color
S	s^i	"Irish" spotting, little white (see text)	Basenji
	s^p	Piebald spotting, up to 80 percent white	Beagle
	s^w	Extreme piebald, small black patches	Bull Terrier, Sealyham
T	*T*	Ticking; flecks or spots in white areas	Some Hounds, Pointers; Dalmatian
	t	No ticking; white areas clear	Most breeds

Source: Condensed from Little (1957).

This is all very well, but most dog breeders will want to know—not just what genes their favorite breed is believed to carry—but what colors and patterns to expect in the pups when dogs of the same color, or of different colors, are mated together. Such expectations for different colors in Cocker Spaniels are shown in Table 2-1 of this book. Others, based on his long experience, were given by Whitney in the latest edition of his well-known book (1972) and by Burns and Fraser (1966).

Readers particularly interested in colors should read Chapters 4 and 5 in Burns and Fraser (1966), if for no other reason than to see that some people do not accept all of Little's genes as gospel. For one of them *(G)*, no genetic evidence was given. For the series of multiple alleles at the *S* locus, some of us would like to see more evidence that four alleles really account for all the variation and why some undetermined number of modifying genes, not necessarily at the *S* locus, might not equally well be responsible for the differing degrees of white spotting from Sealyham Terriers to Basenjis. Irish spotting (so labelled by an earlier investigator who studied it in rats) consists of white spots or streaks in one or more of several areas. These include the muzzle, forehead, chest, feet, belly, and tip of tail.

One difficulty besetting any attempt to analyze white spotting (and other colors too) is that the classifications (as much, little, or less) have to be somewhat arbitrary, and hence are more difficult to make than when dogs are either black or red, solid color or spotted. Accordingly, we should not belittle Little's conclusions. In his useful book, the reader will find the presumed genotypes for many breeds not mentioned in Table 8-2.

According to Burns and Fraser, the greatest puzzle concerning the dog's genes for color is set by those that determine tan, yellow, red, fawn, and golden colors. Anyone will agree who has seen the range in these among Cocker Spaniels that are neither black nor brown (or liver or chocolate). Whitney did not accept Little's classification of black with "flecks"—not ticks. There seems to be disagreement about whether a gene for brindling does or does not belong in the series of alleles at the *E* locus.

From analyses of color in Setters, Cocker Spaniels, bird-dogs, and Pointers in Europe, Winge (1950) developed a genetic interpretation of the basis for color in those breeds. Although it fitted his data well for the breeds that he considered, parts of it have been rejected by other investigators. For one thing, the symbols used by Winge differed from those used earlier by other geneticists. The relationships between them have been nicely shown by Asdell (1966). Winge postulated an epistatic red, capable of producing a red coat in dogs that would otherwise be black or brown, and believed to be carried by those Irish Setters with the deepest red color.

Apart from theories that attempt to explain colors and patterns in all breeds, some good studies have been made to determine the genetic bases for them in different varieties within one breed. An example is that of Warren (1927) with Greyhounds, which was based on data taken from the breed's stud books. As he pointed out, the name Greyhound is really a misnomer, because the breed comprises blacks, brindles, fawns, reds, blues and whites, solid colors, and white spotting—but no "greys." He concluded that the first four of those colors are determined by a series of multiple alleles, that blue is recessive to black, and that white spotting, although recessive to self (solid) coat, could vary all the way down to dogs that were classified as white.

Other studies of coat color in specific breeds are listed in the books by Winge (1950), Little (1957), Burns and Fraser (1966), Whitney (1972), and Wegner (1975). The incompletely dominant gene *M* causing the merle type of dilution of pigment, with serious defects in homozygotes, is considered in Chapter 1. The gray color in Collies, an indicator of a peculiar, lethal type of anemia, is discussed with other genetic abnormalities of the blood in Chapter 10.

9 Nerves and Muscles

There are some hereditary defects in which atrophy or faulty function of muscles is caused by an abnormality in the nervous system. In dogs, a good example (discussed in Chapter 2) is the paralysis of muscles in the thigh which Stockard (1936) found to be caused by a loss of motor neurones in the lumbar region of the spinal cord. Others will be considered in this chapter.

In other kinds of paralysis, the primary defect seems to be in the muscles themselves. In at least one case of malfunction of the muscles, it seems probable that the underlying cause is a defect in the nerves that normally supply those muscles, but the specific lesion has not yet been found. That one case is the "bird-tongue" defect (see Chapters 1 and 6), in which the muscles necessary for swallowing and for movement of the tongue do not function. Let us now consider some genetic defects not mentioned in previous chapters.

Ataxia

My dictionary tells me that locomotor ataxia is "a constitutional unsteadiness in use of legs, arms, etc." The word was used by Bjork et al. (1957) to describe a hereditary disorder in Fox Terriers, and that definition fits perfectly the symptoms seen years later (by me) shortly after their onset in two Kerry Blue Terriers. However, readers must be prepared to meet the disease under other names, such as the "chorea" used by veterinarians in California and the formidable "cerebellar cortical and extrapyramidal nuclear abiotrophy" assigned by veterinarians at Cornell and Harvard to describe that same affliction in Kerry Blues.

The difference between the two short names and the long one can be attributed to the fact that the former are based on clinical symptoms and the latter on histological studies of the lesions in the brain responsible for those symptoms.

In the Swedish Fox Terriers, the ataxia started between 2½ and 4 months of age, when the affected pups began to bark or howl and sought raised places such as a stone or the roof of a kennel. They rubbed their heads against convenient objects and soon showed difficulties in going up or down steps. Bouncing, dancing movements were typical, and locomotion became more difficult. After some months, the symptoms stabilized. The disease was not fatal, but the afflicted animals were usually destroyed before reaching sexual maturity.

Among 91 pups in 23 litters, there were 25 cases of this ataxia. Both sexes were affected, and litters that had ataxic pups traced back to several common progenitors. The condition was clearly caused by an autosomal, recessive gene in the homozygous state.

It is not yet clear whether or not the disease in the Swedish Fox Terriers is the same as that studied by de Lahunta and Averill (1976) in Kerry Blue Terriers. Age at onset was about the same in both breeds (the average being 11 weeks in the Kerry Blues); so was the mode of inheritance. Clinical symptoms differed somewhat in the two breeds. The Kerry Blues did not bark or howl. In that breed, moreover, contrary to the stabilization reached in the Fox Terriers, the disease became progressively worse. By 8 to 10 weeks after the first signs of the disease in Kerry Blue pups, walking was so difficult that the animals frequently fell backward. After 20 weeks, they were unable to stand without support. In autopsies at different ages, progressive degeneration of the cortex of the cerebellum, along with a loss of Purkinje cells, was found in the brains of the afflicted dogs. Comparison of these lesions with those found by Björck et al. (1962) in the Fox Terriers is left to the pathologists.[1]

De Lahunta and Averill (1976) studied 10 ataxic dogs that occurred among 23 pups in five litters at one kennel. That ratio of 13 : 10 is so far from the familiar 3 : 1 that one investigator doubted if the condition could be a simple recessive trait. Further analysis showed that it did qualify for that mode of inheritance (see Chapter 6, last two paragraphs).

The question whether or not hereditary conditions (such as this ataxia) with slightly differing symptoms and lesions in different breeds are really one and the same, or are caused by entirely different genes, is likely to come up often in the future as breeders become more concerned about genetic defects.

---

[1]This is not an error. The Björck of 1962 is the Bjork of 1957. We authors have to follow the printed word in our citations.

(a)

(b)

FIGURE 9-1
Crouching posture and gait in a Weimaraner afflicted with spinal dysraphism. (a) Moving straight ahead. (b) Turning to the left. [From McGrath, 1965.]

Spinal Dysraphism

The name given by McGrath (1965) to peculiar abnormalities of locomotion, studied by him in Weimaraners, is *spinal dysraphism.* Difficulties were usually evident by four to six weeks of age. The abnormalities always included (but not all at the same time!) hopping, a peculiar stance in which the dog crouched as if it were frightened, wide spreading of the hind legs, and over-extension (stretching backward) of the hind legs (Figure 9-1). In young animals, the hind paws became knuckled under, but the symptom disappeared as the dogs matured. These clinical symptoms were attributed to cavities and other lesions found at postmortem examination in the spinal cord. Apparently the severity of the symptoms varied greatly in different dogs.

McGrath's experimental matings included some dogs among both parents and progeny that were classified as showing (either when alive or in postmortem examination) only "subtle signs". Altogether, his seventeen matings yielded 67 dogs that survived long enough for classification. Among these there were 17 normal, 25 clearly affected, and 25 more with subtle signs. No clear relationship was evident between severity of the condition in the parents and the frequency or severity of the syndrome in the progeny. Most of the dogs contributed by breeders for McGrath's study were said (by the donors) to have come from normal parents. However, the condition is not any simple recessive character, for, in one of McGrath's own matings, parents both affected had two normal offspring. From the facts that (1) the syndrome seems to be confined to one breed, (2) that in the experimental matings every affected dog had at least one parent that was affected either clearly or mildly, and (3) the proportion clearly affected was over 37 percent, it seems safe to conclude that the defect is inherited, perhaps incompletely dominant, perhaps polygenic.

Later studies by Draper et al. (1976) suggest that the dysraphism is caused by an incompletely dominant gene that is lethal to some homozygotes at birth or before. Its expression in heterozygotes is so variable that some apparently normal dogs from affected parents can carry the gene without showing it, and, when mated *inter se,* will produce affected offspring. They reported that the abnormality is known in six breeds, but is most frequent in Weimaraners.

Neuronal Abiotrophy

Relax! When I first heard of neuronal abiotrophy, from Dr. Björck, he referred to it as *muscular dystrophy,* and, for most of us, it will be easier to think of the abnormality in those terms than in any others. It has been reported thus far only in the Swedish Lapland breed (which is one kind of Spitz) by Sandefeldt et al. (1973).

Onset of the disease began at five to seven weeks, when weakness of either the front or hind legs was evident. Within a week, all four legs were affected, and by another week the afflicted dog was unable to stand. It lay on its chest, front legs extended forward and hind legs extended backward, with all four paws bent under (Figure 9-2). Some dogs could not move their limbs, but others managed to crawl forward by using muscles closest to the body. Muscles of the trunk were less affected, and the dogs were able to move their heads normally and even to wag their tails. In the legs, the muscles wasted rapidly, and mobility of the joints decreased. Both sexes were affected.

By about two weeks after first symptoms, the disease seemed to have run its full course, and, even in dogs kept to the age of 18 months, there was little further change, if any. Most of the dogs were destroyed at the request of their owners, but 3 (of 10 with reliable histories) died of bronchial pneumonia.

The condition was labelled by Sandefeldt et al. as *neuronal abiotrophy* because of the extensive degenerative changes found on histological examination in the central and peripheral nervous systems. These are described in detail in the report cited.

In 12 litters that contained at least one affected puppy, the ratio of normal to afflicted dogs was 43 : 19. The muscle dystrophy is clearly a simple recessive, autosomal character, with complete penetrance. Although some dogs with the disease were kept alive for 18 months, it is obvious that none could have survived long in the wild, so the defect can be added to the list of hereditary lethal characters in the species.

Paralysis (Myelopathy) in Afghan Hounds

Myelin is a white fatty substance that forms a sheath around some nerve fibres. Several veterinarians have reported paralysis of the limbs

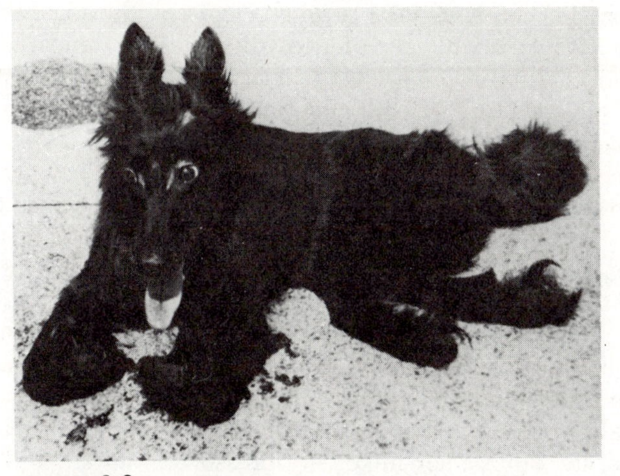

FIGURE 9-2
A Swedish Lapland dog with muscular dystrophy. All legs were paralyzed, forelegs extended frontward, hind legs as shown, muscles of all legs atrophied. [From Sandefeldt et al., 1973.]

in Afghans which is associated with necrosis of the myelin in the spinal cord. The region affected is that of the thoracic vertebrae, but it may extend farther in severe cases.

According to Averill and Bronson (1977), clinical signs of the condition are recognizable at 3 to 8 months of age, the average in their 11 cases being 5.7 months. Weakness in the hind legs and resultant unsteadiness (ataxia) progressed within a week to nearly complete paralysis of those limbs, but knee jerks and ankle jerks persisted. Front legs became affected a week or two later than the hind ones, but their weakness progressed rapidly in most cases so that the dogs could not stand up.

Of the cases described by Cockrell et al. (1973), four were in one litter and two others were also siblings. This, along with the incidence in one breed, suggests that the condition is hereditary. Final proof was adduced by Averill and Bronson, who concluded from the study of four litters (13 affected animals descended from one female) that the defect is a simple, autosomal, recessive trait. As none of the dogs recovered, it is obviously a lethal character.

Epilepsy

Epilepsy is known in several species of mammals and birds. It has been reported in so many breeds of dogs that none should be considered exempt. It seems probable that, while epileptic seizures can result sometimes from injury to the brain and other causes, most cases are hereditary defects. A general review of epilepsy in dogs was given by Eberhart (1959). In my experience, the questions most often asked about it by dog breeders are (1) Is it hereditary? (2) Is it safe to breed from normal relatives of epileptics? and (3) How can it be eliminated from a kennel?

Critical genetic evidence that epilepsy is hereditary is difficult to obtain in the dog because manifestation of the condition is uncommon in pups younger than a year, but some dogs may not show it until past six years of age. This makes results of experimental matings both difficult to interpret and unrewarding. Burns and Fraser (1966) cited some evidence that epilepsy might be a simple recessive character. Falco et al. (1974), who studied the pedigrees and records of Alsatians (German Shepherds), concluded that the genetic basis is complex, and Bielfelt et al. (1971) postulated genes at two loci, one autosomal and recessive, the other a sex-linked suppressor. So far as the dog breeder is concerned, the important thing about these three reports is their agreement that epilepsy in the dog is a hereditary defect.

The most conclusive genetic study of epilepsy in a vertebrate is probably that of Crawford (1970) in the domestic fowl. The seizures in his epileptic chicks were similar to those of grand mal epilepsy in man. So also (as later studies revealed) were the chicks' electroencephalograms and their reactions to anticonvulsant drugs. Best of all, for genetic studies, chicks susceptible to epilepsy showed the condition either at hatching or (when suitably excited) within 24 hours. They also lived to reproduce.

With this ideal model, Crawford hatched 2,282 chicks for the usual Mendelian analyses and found that the condition was a simple, autosomal, recessive trait with incomplete penetrance in chicks from heterozygous dams. Epileptics mated *inter se* yielded (in offspring that lived 24 hours) only epileptic chicks. It will be recognized that genetic tests on any such scale are not likely to be attempted in dogs. The findings are not directly applicable to other species, and the mode of inheritance in dogs may be more complex, but Crawford's studies prove beyond any question that epilepsy can be a genetic defect.

Returning now to the other questions about epilepsy raised earlier, it is *not* safe to breed from normal relatives of epileptic dogs, and the best way to eliminate the disease is not to breed from affected dogs, their litter-mates, their parents, or (if there are any) their offspring.

In some breeds, epilepsy has been so prevalent (most likely from inbreeding) that the breed societies have become alarmed and have sought veterinary aid to reduce it. Examples are the Keeshond in Britain (Croft, 1968) and the Tervueren Shepherd, a variety of the Belgian Sheep Dog, in Belgium (Van der Velden, 1968). Croft found the electroencephalograph useful in detecting Keeshonds susceptible to epilepsy, before they had had any fits. Its use for screening dogs that might be used for breeding has been recommended. It seems far from clear, however, that the electroencephalograph can detect every dog that has the genotype for epilepsy. Even if it did so, and if all such dogs were excluded from reproduction, that screening and restriction would be fully effective only if epilepsy were caused by a dominant gene. The evidence suggests that it is more likely to be recessive. In that case, even if all potentially epileptic dogs were excluded from breeding, the gene would still be carried along in the normal but heterozygous individuals, which (thus far) can apparently not be identified from their encephalograms. Some day they may be, but until that day comes, it would seem safer to avoid using epileptic dogs and their near relatives for breeding.

Sex-linked Myopathy

The peculiar disease of the muscles which I call *sex-linked myopathy* was studied by Wentink et al. (1972) in Irish Terriers in the Netherlands. Symptoms began after 8 weeks of age, when affected pups showed difficulty in swallowing, which was accompanied by enlarged tongues. By 13 weeks, the dogs walked stiffly and were unable to jump or to lift themselves on their hind legs. Their backs became curved, and cheeks were moist and dirty from feed residues. Their mouths could be opened only partly. Enlargement of the tongue was conspicuous. In spite of all these disorders, the afflicted pups did not show pain, but slight struggling when they were handled caused immediate cyanosis. The muscles underwent progressive atrophy and became difficult to relax.

On electro-myographic examination, the muscles showed "high-frequency discharge . . . in a dive-bomber pattern," and, in dogs killed at ages up to six months, the muscles were found at necropsy to be pale, with yellowish white streaks. Microscopical and chemical studies of the diseased muscles led the investigators to conclude that the disease was a primary myopathy; i.e., originated in the muscles. The litter of 8 pups that brought this condition to the attention of Wentink et al. included five affected, all of which were males, and three normal (2♂♂, 1♀). The breeder had excellent records covering 9 previous litters, in 5 of which there had been pups afflicted with the myopathy (Figure 9-3). Altogether, in the 10 litters there were

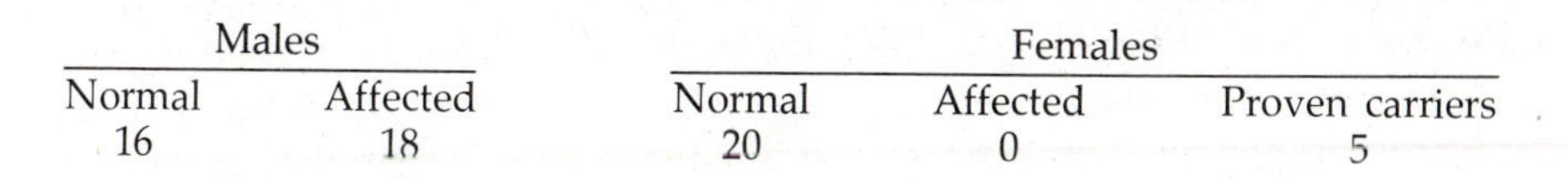

Males		Females		
Normal	Affected	Normal	Affected	Proven carriers
16	18	20	0	5

With commendable caution, Wentink and his associates referred to the myopathy as showing "possible recessive X-linked inheritance,"

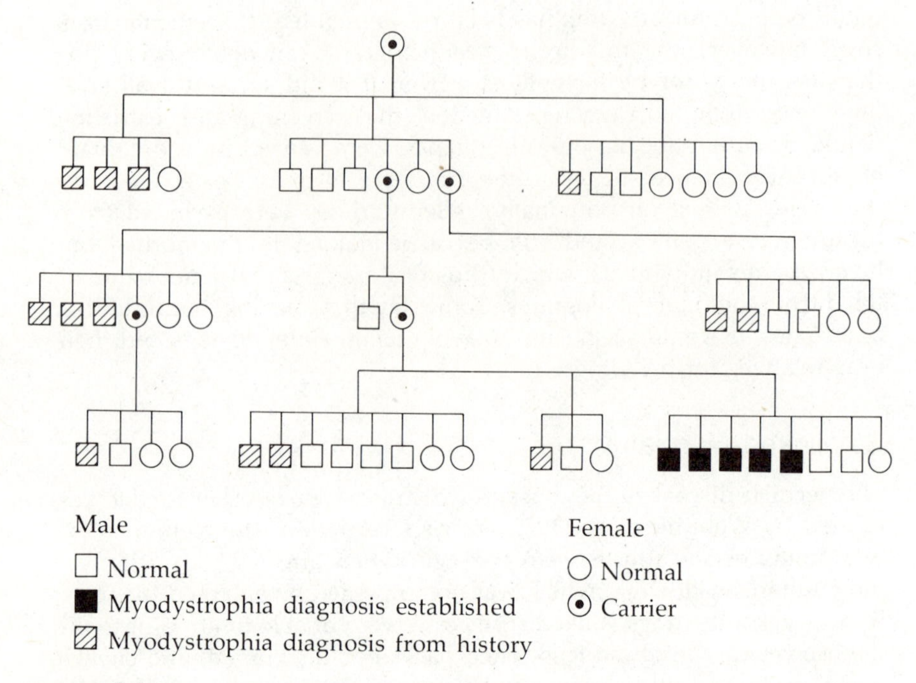

FIGURE 9-3

Pedigree of the Irish Terriers afflicted with sex-linked muscular dystrophy. The symbols show 5 males studied, 13 more diagnosed as dystrophic from descriptions by the breeder, and 5 females proven to be carriers of the causative gene. [After Wentink et al., 1972.]

but, from the records depicted in Figure 9-3, as just summarized there can be little doubt that the causative gene is on the sex chromosome. It becomes the fourth to attain that distinction.

Scottie Cramp

The condition sometimes designated as *Scottie cramp, Scotch cramp,* or *recurrent tetany* owes its name to the facts that it was first recognized in Scottish Terriers as a specific disease, and that, although there are several references to the condition in the literature, it seems to be confined almost entirely to that breed. Symptoms are not evident when the dog is resting, or during light exercise, but are precipitated by excitement or strenuous exertion. They begin with arching of the back, followed by an abnormal gait (Figure 9-4). Walking is difficult because the hind legs become bent and the front legs extended. Some affected dogs make progress by hopping, but others are unable to walk. Some fall on one side and remain for some time unable to stand.

Signs of Scottie cramp are sometimes evident as early as six weeks of age, and most of those affected show the symptoms during their

FIGURE 9-4
A dog stricken with Scottie cramp while walking. Arching of the back, immobile hind legs, and stiff front legs are typical. [From Meyers et al., 1970.]

first year. The condition is believed to have no adverse effects on viability. Meyers et al. (1969) found that symptoms could be allayed by diazepam, a drug commonly used to relax skeletal muscles, or by tranquilisers. They concluded from extensive studies that Scottie cramp is caused by some defect in the central nervous system.

Indications that the condition is hereditary date from the studies of Klarenbeek et al. (1942), who observed five cases in related animals at one kennel. The fact that it seems thus far to be restricted to one breed also suggests that it is genetic in origin. Conclusive evidence on that score was provided by Meyers et al. (1970), who, from reliable reports from breeders and some matings at Washington State University, determined the frequency of normal and affected dogs in seventeen litters that came from normal parents but included at least one affected animal. Altogether, these litters provided 83 pups that lived at least to 18 months. Among them, the ratio of normal : affected was 59 : 24, a close fit to expectations for a simple recessive trait. The critical test of mating affected dogs *inter se* yielded, in three litters, 10 dogs that lived over six weeks. All 10 showed Scottie cramp. In the 56 affected dogs, both sexes occurred in approximately equal numbers, a good indication that the causative gene is autosomal.

Myotonia

Three cases of myotonia studied by Wentinck et al. (1974) in one family of Chow Chows provided some evidence that the condition is probably a simple, autosomal, recessive trait.

Myotonia is defined as the active contraction of a muscle which persists after stimulation or voluntary effort has ceased. The first symptoms became noticeable at three months of age, when stiffness of gait and excessive salivation were noticed. After a period of inactivity, the front legs were spread apart and the hind legs held close together. As the dog walked, with forelegs fully extended, the weight of the body was shifted from side to side, causing a swinging gait. After 10 to 20 seconds of walking, the legs moved normally, but lifting the dog caused spasm of all voluntary muscles. Other abnormal reactions of the muscles were found in laboratory tests.

Myotonia was not lethal in these Chow Chows. Contrary to what happens in many hereditary disorders, the symptoms became less severe after the first year.

In two litters of Chow Chows from the same normal parents, there were three myotonic pups—a female in one litter and two males in another. Five litter-mates were presumably normal, but not all were examined. Another Chow Chow with this condition was studied by Griffith and Duncan (1973), but their other cases—in a Fox Terrier, a West Highland White Terrier, and an Alsatian (German Shepherd)—showed that myotonia is not restricted to any one breed.

Globoid-cell Leukodystrophy

Hirth and Nielsen (1967) gave the name *globoid-cell leukodystrophy* to an abnormality studied by them which appeared to be so similar to Krabbe's disease in man that that two might be considered identical. In both species, large, lightly stained, globoid cells are found widely distributed in the white matter of the brain. The condition in man is recognized as being "familial," and in the Cairn Terriers of Hirth and Nielsen it was transmitted as a simple, recessive, autosomal character (Figure 9-5).

In those Cairn Terriers, the first symptoms (which became evident at four months of age) could be described as incoordination and stiffness in the hind legs. The afflicted pup had difficulty in climbing stairs. The hind legs were kept far apart, as if to maintain balance. When the pup attempted to run after a thrown ball, it would suddenly fall over because the hind legs had collapsed. By six months of age, one such dog had difficulty in standing, and showed tremors, with twitching of the head. Because of its distress, this one was destroyed. Another was euthanized at an earlier age in order to study the early lesions of the disease.

Hirth and Nielsen refer to other cases of the same disease in Cairn Terriers and in West Highland White Terriers, but, as always, we must remember that mutations can occur in any breed—or person.

For this addition to our knowledge of the dog's bad genes, we are all indebted to the owner of these dogs. By the time that two litters had

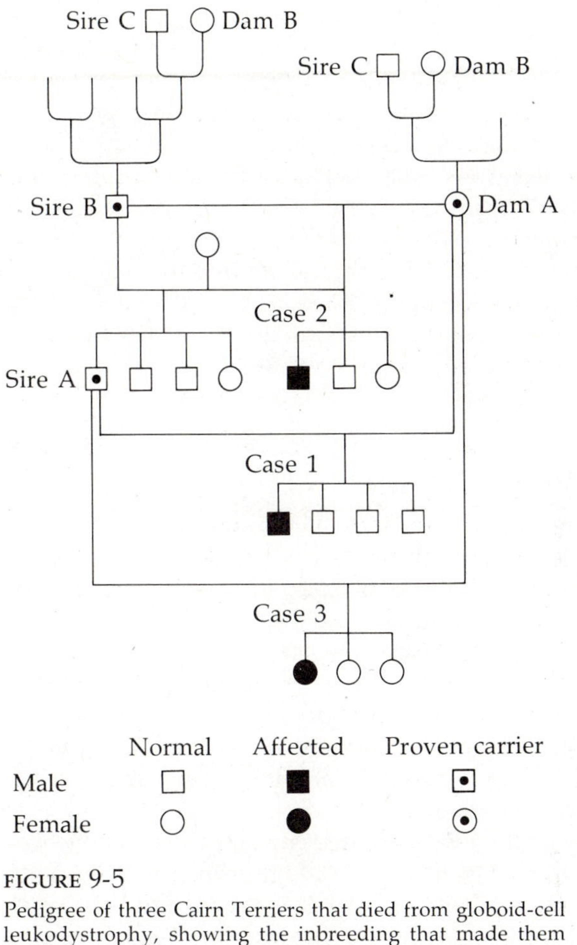

FIGURE 9-5
Pedigree of three Cairn Terriers that died from globoid-cell leukodystrophy, showing the inbreeding that made them homozygous for the causative lethal gene. [From Hirth and Nielsen, 1967.]

come from the same parents, each with an afflicted pup, she knew that the condition was probably inherited, and that one or both of the parents must have carried the causative gene. When asked to repeat the mating in order to provide additional genetic evidence and a puppy for special laboratory examinations, she agreed. The third litter provided an affected female, which proved that the condition is not sex-linked. It also provided a diseased brain for the histological studies, which revealed the lesions characteristic of Krabbe's disease in man.

Leukodystrophy in Dalmatians

A somewhat similar leukodystrophy was studied by Bjerkås (1977) in Dalmatians in Norway. It differed from that just described in causing cavities in the white matter (myelin) of the cerebral hemispheres of the brain. The optic nerves were frequently affected, and in 2 of the 7 cases examined there were also lesions in the spinal cord.

Onset, at 3 to 6 months of age, was recognizable by failing vision and ataxia that began in the hind legs, but later involved all four. Most of the affected dogs were killed within 4 months after onset of the disease. By that time some dogs had paralysis of the hind legs. Both males and females were affected.

Counting 4 cases with clinical symptoms that were not verified by post-mortem examinations, Bjerkås had a ratio of 32 normal : 11 affected in seven litters all descended from one male. A better fit to the 3 : 1 ratio expected for a simple recessive, autosomal trait could not be attained.

Swimmers

In doggy and veterinary jargon, the term *swimmers* is really more descriptive of a condition found in newborn puppies than the other term sometimes used for it: the flat-pup syndrome. The pups afflicted do lie flat on their bellies, but it is the movements of their legs as they try to propel themselves forward that earn for them their very apt designation as swimmers. With special care, many of them manage to get their legs where they should be and eventually recover completely from their initial handicap. Swimmers are said to occur more frequently in small breeds than in big ones, and it has been suggested that the condition may be hereditary.

Palmer et al. (1973) reported the occurrence of 30 swimmers among 96 puppies in eleven litters of highly inbred Irish Setters. Although their numbers of 59 normal : 30 affected (and 7 not classified) fit a 2 : 1 ratio better than the expected 3 : 1, the excess of affected pups probably resulted from the fact that breeders keep track of the abnormal pups better than of the normal ones.[2] At any rate, there can be no doubt that the condition is hereditary, recessive, autosomal, and attributable to a single gene in the homozygous state. Under natural conditions, it would be lethal soon after birth. Palmer et al. studied 18 affected pups and managed, with special care, to raise them and even to keep one alive for twenty-five weeks. Others were destroyed at various ages for histological examinations.

It is clear that the condition in these dogs was not the same as that in most swimmers. The afflicted Irish Setters were blind; most swimmers are not. None of the Setters recovered; other swimmers do. Addi-

---

[2]Weinberg's "sib method," by which Palmer et al. sought to correct for the excess of affected pups, is applicable only when there is a classification for every pup in the litter. Of their 11 litters, 4 did not so qualify.

tional manifestations of disease in the Setters included tremor, uncontrolled movements, fits, and nystagmus (involuntary rapid movements of the eyeball). Histological studies showed in the older dogs abnormalities in the cerebellum, particularly a loss of Purkinje cells. Palmer and his associates called the condition *quadriplegia* (paralysis of all four limbs), but the descriptions given show that the legs were only partially paralyzed.

10 Blood and Cardiovascular Abnormalities

Many hereditary defects in the blood of dogs have been identified and studied in the past thirty years, and this section will cover briefly what is known about them. Some of these diseases are not easily recognized or identified, so dog breeders can only report suspicious or unusual symptoms to their veterinarians and arrange through them (when necessary) for final diagnoses in laboratories where specialists and facilities are available. It is helpful, however, if the breeder knows what kinds of defects there are and what symptoms merit further investigation.

Abnormal Clotting of the Blood

Clotting of the blood, once thought to be a simple process, is now known to be very complex. Specialists in that field have identified 13 "factors" (proteins) believed to be important or essential for complete, normal clotting, and these are identified as factors I to XIII. Proponents of the "cascade" or "waterfall" theory of blood clotting hold that each of these permits one further step in the complicated process, and that a deficiency of any one factor interferes with clotting in the "intrinsic" (intravascular) system of the blood. Fortunately, coagulation is aided in that event by the "extrinsic" clotting system; i.e., substances from other tissues of the body. The extrinsic system serves to enhance the sequential events of the intrinsic system. Defects of the intrinsic system significantly prolong the clotting process. For example, in dogs with a deficiency of factor VIII, clotting takes at least five or six times as long as is normal, and the dog has either hemophilia A, or (less commonly) von Willebrand's disease.

It is beyond the scope of this book to delve into the roles of 13 factors that cooperate to cause clotting of the blood. Those who wish to do so should consult the book by Hall (1972), who outlines the laboratory tests for identification of those factors, their interactions, and the effects of various deficiencies. Some more recent studies are cited in the following paragraphs.

Deficiencies in several factors in the clotting process are now recognized as being hereditary in dogs. Description of these genetic defects is here restricted mostly to clinical symptoms which may alert the dog owner to the possibility that something is wrong with the blood. The laboratory tests required for exact diagnoses are described in various books written for veterinarians and hematologists.

Factor VII

A deficiency of factor VII[1] was first identified in Beagles in Canada by Mustard et al. (1962). Subsequently it has been found in two large colonies of Beagles in Britain and (in that same breed) in the United States. All of this does not mean that the defect is restricted to Beagles. It is more likely that the laboratory tests necessary for its recognition are made only infrequently with other breeds. Dodds (1974) found it in Alaskan Malamutes.

Clinical symptoms of a deficiency of factor VII in Beagles are usually so slight that they may be overlooked. Bruises incurred from fighting may result in hematomas (subcutaneous swellings full of blood). Although such visible effects are mild, a deficiency of factor VII may cause trouble in times of stress. Spurling et al. (1974b) cite three such cases that followed parturition. It is also possible that effects of the deficiency could be more serious in other breeds than in Beagles.

From studies of pedigrees, Rowsell (1963) thought that a deficiency of factor VII is a simple, autosomal recessive trait. That opinion is supported by an extensive pedigree of the defect in the Beagle colony of Allen and Hanburys in England, and is confirmed by three litters in that pedigree from parents both deficient in factor VII. These contained, altogether, 15 pups, every one of which had the same deficiency as the parents (Spurling et al., 1972).

However, as with many other genetic variations in constituents of the blood, heterozygotes show (in appropriate laboratory tests) some influence of both the recessive and dominant allele. The two are codominant. It is not possible to recognize the heterozygotes from their appearance, but it is possible to do so in many cases by laboratory tests to determine the levels of factor VII in the blood.

---

[1]You can call the deficiency hypoproconvertinemia if you wish.

Although factor VII apparently poses no problems for most kennels, the deficiency of it can be serious in the large colonies of Beagles maintained to supply dogs to laboratories for various kinds of physiological and pharmacological studies. According to Dodds et al. (1967), factor VII is used up quickly in the body, and the level of its activity varies greatly. This, in turn, causes variation in its interactions with other factors that influence coagulation of the blood. Accordingly, it is now customary for laboratories using Beagles for experimental tests to screen out dogs showing a deficiency of factor VII.

Obviously, the best procedure would be to eliminate the heterozygotes from the colony that produces the dogs. Hitherto that has not been considered feasible because, by the Thrombotest method used to measure the level of factor VII activity, the range of values determined for the heterozygotes considerably overlapped that for normal dogs. Recently, Spurling et al. (1974a) have devised a more sensitive modification of the Thrombotest method which differentiates the two genotypes more accurately. It thus reduces the overlap of the two ranges so much that almost all of the heterozygotes can be identified, and comparatively few of the homozygous normal dogs would be erroneously eliminated as suspect (Figure 10-1).

Factor VIII

Factor VIII, also called *antihemophilic factor* or *antihemophilic globulin,* was discussed in Chapter 4. When it is deficient in dogs, they show all the symptoms of true hemophilia, or hemophilia A, the most common kind of hemophilia in man. In both species, the causative gene is recessive and sex-linked. The same defect is also sex-linked in the horse. Hemophilia is the most common bleeding disease in dogs. According to Dodds (1974), it accounts for about 80 percent of such cases, and has been recognized in almost every pure breed as well as in mongrels.

Factor IX

Factor IX, also called *Christmas factor,* or *plasma thromboplastin component,* has also been considered in Chapter 4. A deficiency of it causes hemophilia B, or Christmas disease, which is equally as severe as hemophilia A, but much less common. The causative gene was the second one to be identified in the X chromosome.

Factor X

As with the foregoing three factors, a deficiency of this one (also called the *Stuart-Prower factor*) was first known in man, but Dodds (1973) recognized it in a large family of Cocker Spaniels. In dogs, as in man, the defect is hereditary.

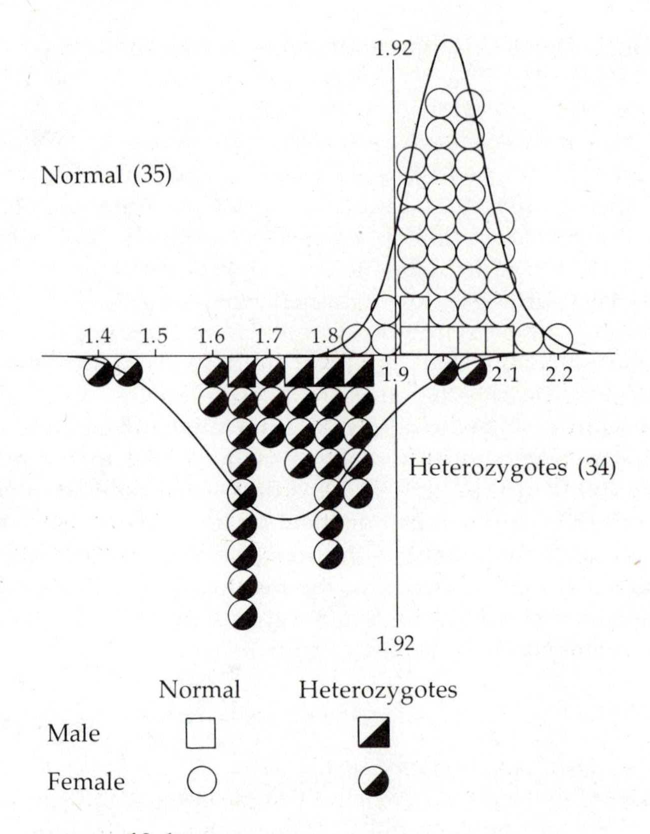

FIGURE 10-1
Frequency distributions for factor VII activity assayed in Beagles and expressed logarithmically. Heterozygotes are inverted to show better the slight overlapping of the two distributions. The "discrimination threshold" (1.92) is equivalent to 83.5 percent factor VII activity. For 64 of the 69 dogs, the test would have identified the genotype correctly. [After Spurling et al., 1974a.]

In Cocker Spaniels, deficiency of factor X causes unusually high mortality of puppies. These are stillborn or die a few days after birth as a result of massive internal hemorrhages in the thorax or abdomen. Clinical symptoms seen at later ages in dogs that survive the critical first-10-day period include prolonged bleeding after surgery (even after tail-docking), bleeding fore or aft (from the nose or from the rectum), and spitting blood. Newborn pups may show protracted bleeding from the umbilicus. One affected dog that lived 11 years had recurrent bleeding from the gums. Several hours before death, some pups showed signs of hemorrhage under the skin covering the abdominal wall (Dodds, 1973).

In 14 litters from controlled matings made to study the genetic

TABLE 10-1 Deficiency of Factor X in Test-matings

Series	Parents	Litters	Progeny			
			Died, other causes	Normal[a]	Deficient in Factor X	
					Mild	Severe
1	Both normal	5	4	24	0	0
2	One normal, one mildly affected	6	10	8	7	8
3	Both mildly affected	3	4	1	2	8

[a]Proven normal by clotting tests.
Source: Data of Dodds (1973).

basis for the deficiency, Dodds obtained 76 pups. Because the trait was clearly autosomal, her separate results for each sex can be combined. Lest any dog breeder might think that the do-it-yourself principle can be applied to tests for recognition of a deficiency of factor X, it is worth recording that Dodds measured that deficiency (in percentage) as the extent to which individuals attained normal clotting in a series of tests with several reagents. By those tests, she measured the recalcification times, thrombin times, partial thromboplastin times, one-stage prothrombin times, and Russell's viper-venom times, and also made specific assays for seven different factors in the clotting mechanism. Out of all this, the dogs were rated as having a mild deficiency if their factor X activity was 25 to 65 percent of normal, and a severe deficiency when it was lower than 25 percent of normal. Pups that died early from massive hemorrhages were classified as severely affected; so were those stillborn and found at autopsy to have been bleeding internally.

The results in Series 2 of Dodd's matings (Table 10-1) suggest that the deficiency of factor X is caused by an incompletely dominant gene. The range of its expression in heterozygotes runs all the way from lethal hemorrhages soon after birth to mild, protracted bleeding in older dogs.

One might expect that puppies homozygous for the causative gene would be affected much worse than the heterozygotes (as in hairless dogs and merles), and there is evidence in the Dodds data that this is so with the deficiency of factor X. From the fact that in Series 3 (Table 10-1), where homozygotes could occur, the average size of litters (5) was little different from the 5.6 in the other two series, it seems probable that any prenatal mortality does not strike the homozygotes at early stages of gestation. On the other hand, the proportion of affected pups that had a severe deficiency (some of which were stillborn) was much higher in Series 3 than in Series 2, where no homozygotes could occur.

Steps taken by the American Spaniel Club to eliminate this undesirable mutation are considered in Chapter 13. There is nothing to prevent that same mutation from occurring in some other breed.

Von Willebrand's Disease (VWD)

Von Willebrand's disease is hereditary in man, swine, and dogs. The disease results from more than one abnormality in the blood-clotting system, including reduction of the activity of factor VIII in that system. It thus delays clotting. It has been recognized in German Shepherds, Miniature Schnauzers, Golden Retrievers, Doberman Pinschers, and Scottish Terriers. Isolated cases have also been studied in other breeds (Dodds, 1974, 1976).

The most frequent symptoms of VWD include formation of hematomas, intermittent lameness (from bleeding into and around the joints) and bloody discharges from body openings. Nose-bleeds are particularly common in Scottish Terriers. Oddly enough, Dodds (1975) found that these clinical symptoms became gradually less severe as the dogs got older, during pregnancy, and after repeated pregnancies.

Recognition of von Willebrand's disease and its differentiation from other defects in blood clotting are tasks for specialists in diseases of the blood. It was first identified by Dodds (1970) in a family of German Shepherd Dogs. Results of her extensive subsequent studies to determine its frequency in experimental matings are summarized in Table 10-2. Because they showed that the defect is autosomal, her separate data for the two sexes are combined in that table.

As the matings of normal × normal produced no affected progeny, while every dog with the disease had at least one affected parent, von Willebrand's disease is clearly caused by a dominant gene. When one parent was affected and the other not, the ratio in the living pups of 25 : 22 fitted closely that expected (1 : 1) for a dominant gene. From other analyses, it seems probable that the gene causing von Willebrand's disease in dogs is lethal to homozygotes, most of which (if not all) are stillborn.

Because both sexes can be affected, it is necessary to distinguish between hemophilia A and von Willebrand's disease in males having a deficiency of factor VIII, unless there is clear genetic evidence that their disease is sex-linked. Specific diagnostic tests are required to make that distinction. If bleeder males have any bleeder sisters, their trouble is not likely to be hemophilia A or B.

Among affected surviving dogs (all of which should be heterozygous), there is considerable variation in the severity of the disease. Dodds found that most of them had primary bleeding times (before clotting occurred) exceeding 10 minutes, instead of the 2 to 5 minutes in normal dogs. Because their laboratory tests were less abnormal than

TABLE 10-2 Experimental Matings Showing Inheritance of Von Willebrand's Disease

Parents	Litters	Progeny			
		Stillborn	Live, Affected	Live, Normal	Total
Both normal	7	2	0	38	40
Normal × von Willebrand	9	5	25	22	52
Both von Willebrand	6	17	21	14	52

Source: Data of Dodds (1975).

others, a few were classified as "incomplete," but some of these proved later to be unusually susceptible to some other pathological conditions.

It seems probable that expression of the dominant gene causing this disease is so variable because of modifying genes that suppress its effects in some dogs. As such modifiers would vary not only in littermates within one breed, but more widely in different breeds, von Willebrand's disease could be more serious in some breeds (and less serious in others) than Dodds found it to be in German Shepherds. For example, in dogs that undergo what is euphemistically termed "cosmetic surgery" (ear-cropping, tail-docking), bleeding from VWD may be severe (even fatal) at an earlier age in some breeds than in others. Affected Miniature Schnauzers and Doberman Pinschers manifest a severe form of VWD at earlier ages than do affected German Shepherds and Golden Retrievers (Dodds, personal communication, 1976).

Factor XI (PTA)

The abbreviation PTA, by which factor XI is known among specialists in blood clotting, stands for *plasma thromboplastin antecedent.* Fortunately, a deficiency of it seems to be comparatively rare in dogs, but it occurs also in man and in cattle. Unlike VWD, which causes bleeding during surgical procedures, a deficiency of factor XI causes hemorrhage within 24 hours *after* surgery. Dodds (1974) encountered it in a Springer Spaniel bitch that experienced nearly fatal hemorrhage 24 hours after being spayed.

Earlier the dog had had a litter from which three pups were available for quantitative assays of factor XI. In all of these (2♂♂, 1♀), the levels of factor XI were only from 23 to 40 percent of normal. This showed that they were heterozygous for the deficiency—that it is hereditary and autosomal. Presumably the mother (at levels of 3 to 10 percent) was homozygous.

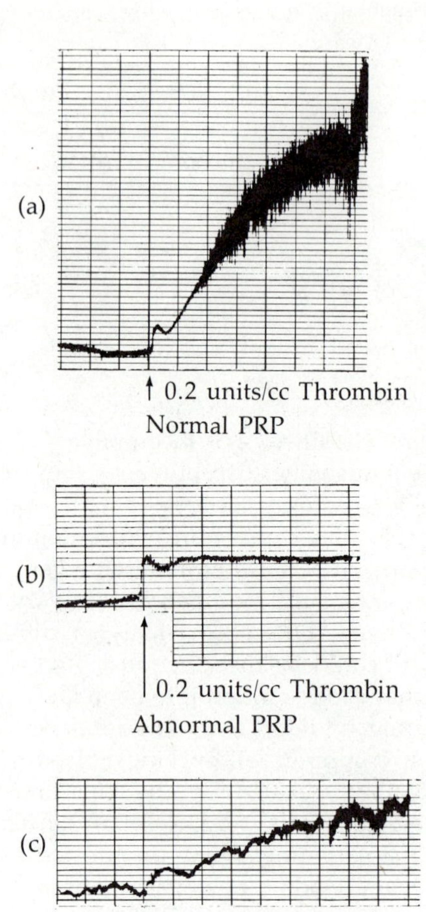

FIGURE 10-2

Thrombasthenia in Otter Hounds. Tests with thrombin to see if the blood platelets aggregate properly show differing responses in (a) normal dogs, (b) those afflicted with thrombasthenia, and (c) heterozygotes. Detection of these last facilitated elimination of the defect. [Courtesy of W. Jean Dodds, Division of Laboratories and Research, New York State Department of Health, Albany, New York.]

Defects in Blood Platelets

The little cells called *blood platelets* or *thrombocytes* normally form a first line of defence against excessive bleeding whenever a blood vessel is ruptured. There are several abnormalities attributable to failure of the platelets to function as they should. One of these, termed *thrombasthenia,* has been known for some years in Otter Hounds, but, as a result of extensive testing and selective breeding, has now been practically eliminated from show stock of that breed (Dodds, 1976). The comparatively rapid progress in attaining that objective was possible because the disorder is apparently caused by a codominant, autosomal gene, and (see Figure 10-2) both homozygous and heterozygous animals can be identified by appropriate tests (Dodds, 1974).

Another hereditary malfunction of the platelets is known in Basset Hounds (Dodds, 1974). It affects both sexes, but the genetic basis for it is not yet clear. As with the thrombasthenia in Otter Hounds, it causes bleeding from mucous membranes, but more severe bleeding after injuries or surgery.

Seven diseases that cause prolonged bleeding have been described in the foregoing pages. For other defects in blood clotting, not all of which are inherited, readers should consult the useful reviews of the subject by Hall (1972) and Dodds (1974, 1976). Obviously, whenever some hereditary defect in the clotting mechanism is suspected, the dog owner should send out an SOS to his (or her) veterinarian.

Other Genetic Defects of the Blood

Cyclic Neutropenia in Gray Collies

The words *cyclic neutropenia* mean simply periodic declines in the numbers of certain blood cells known as *neutrophiles.* They are so called because they can be stained with dyes that are neither acid, nor basic, but neutral. The neutrophiles belong to the class of leukocytes called *polymorphonuclears* because their nuclei vary in size and shape. They are also called *granulocytes* because they are cells containing stainable granules. More serious for the gray Collies, neutrophiles are also phagocytes; i.e., they engulf and destroy invading bacteria; they are thus an important part of the defences against pathogenic bacteria.

As a result, when the neutrophiles decrease in number, resistance to infections is lowered. That happens in dogs affected with cyclic neutropenia at intervals of 10 to 12 days, and each such episode lasts two to three days. During that time, the various clinical symptoms noted either by Cheville (1968) or Lund (1970) included loss of appetite, lameness (from arthritic joints), fever, inflammation of the gums, diarrhea, and infections. Differential blood counts showed that during an episode of neutropenia the neutrophiles may disappear from the

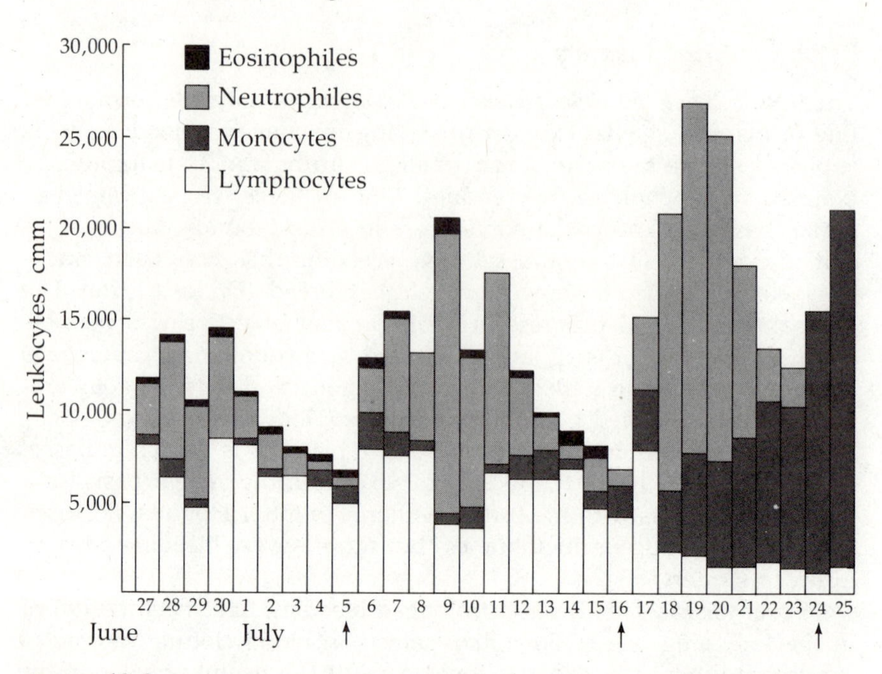

FIGURE 10-3
Fluctuations of four kinds of white blood cells during 29 days that included three episodes of neutropenia (arrows). In the last of these, disappearance of the neutrophiles (gray) and increase of the monocytes was followed by death. [After Cheville, 1968.]

blood entirely or be reduced to abnormally low levels (Figure 10-3). They do so because the bone marrow fails to produce new ones at that time. After each episode, the neutrophiles increase to abnormally high levels before starting the next decline.

Mortality

Most of these gray Collies die within six months of birth from severe infections (Figure 10-4). Ford (1969), who had studied gray Collies for 13 years, found that, among 92 grays born during that period, 67 percent died during the first week, only 6.5 percent lived longer than seven months, and only two dogs lived a whole year. Lund et al. (1970) were able to prevent much of the usual mortality in the first week by raising the gray pups by hand. With antibiotics and other special treatment during episodes of neutropenia, they raised some of them to adulthood, and two even lived for 2½ to 3 years. Such adults were smaller than normal Collies and slower to reach sexual maturity.

Genetics

In Ford's 39 litters that contained one or more gray pups, there was a ratio of 194 not gray : 92 gray. None of the parents was gray. Most

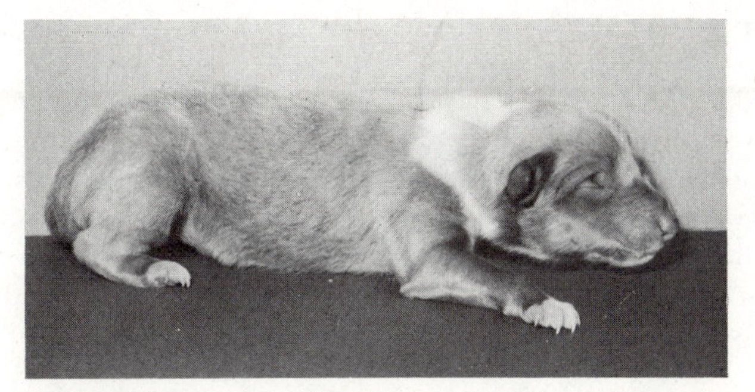

FIGURE 10-4
A sick gray Collie pup during an episode of neutropenia. [Courtesy of N. F. Cheville.]

of them were sable, as were almost 90 percent of the pups not gray. Clearly, this kind of gray color is a simple recessive character. As is usual when the data are restricted to litters containing at least one pup of a recessive type, there was an excess of gray pups above the 25 percent expected, but Ford's correction for that left a good fit of observed numbers to the 3 : 1 ratio expected.

Similar data from breeders' records of 28 litters compiled by Lund et al. (1970) confirmed Ford's finding and showed also that the causative gene is autosomal. Among their 207 pups in litters that contained at least one gray, the proportion of grays was actually 31.4 percent before statistical corrections.

Because the gray mutation was known to Collie breeders for many years before its association with neutropenia was recognized, the latter condition might be considered a pleiotropic effect of the gene causing gray color. However, since all grays (of this type) have the defect of the blood, it is simpler to view the two conditions as resulting primarily from a mutation that interferes with the formation of blood cells. Its effect on color, although recognized first, really may be secondary. For those of us looking for genetic indicators of ability to resist disease, the gray Collies are of special interest because their color is an infallible indicator of acute susceptibility to a specific disease.

Grays, Silvers, and Others

The gray Collies considered above were known to Ford as "grays, silvers or silver-grays." They were said to be "very light silvery gray . . . almost white at times." In addition to these, Ford (1969) described a dominant slate gray that appeared in litters only when one parent showed that same color. These dominant grays were fully normal in viability and reproduction. So also were recessive maltese grays described as being of a darker and deeper mouse-gray color. Evidently Collie breeders cannot condemn all grays but must learn to identify three different phenotypes. This is important, for Ford calculated that about 32 percent of the Collies in the United States could carry the gene for lethal gray.

Cheville's gray Collies were said to range in color "from dark pewter-gray to silver." Lund et al. reported that the effect of the lethal mutation on coat color was to dilute sable to colors that ranged from light, silver gray to dilute beige. It also diluted brown in dogs that would otherwise have been tricolor.

Cyclic neutropenia also occurs in man, but in our species the interval between episodes is about twice as long as in the dog, or more.

Hemolytic Anemia

There are different kinds of hemolytic anemia. In all of them the red corpuscles are reduced in number and some are characterized by the appearance in the blood of small, round erythrocytes, called *spherocytes*. There is another kind, which does not show such spherocytes, and which was therefore described as non-spherocytic hemolytic anemia by Ewing (1969), who studied several cases of it in closely related Basenjis (Figure 10-5). Three other cases were reported in that same breed by Tasker et al. (1969). It is now known to be widespread in Basenjis and to occur also in Beagles (Prasse et al., 1975).

Anemias of any kind are not easily recognized by clinical symptoms, and diagnosis of the specific kind can be made only by laboratory examination of the blood. With that, according to Brown and Teng (1975), hereditary hemolytic anemia can be diagnosed by eight weeks of age. Otherwise, apart from signs of weakness that only an experienced dog watcher might notice, the most reliable visible symptoms of this kind of anemia seem to be pallor of the mucous membranes and (for those who learn how to detect it) enlargement of the spleen in adult dogs.

Ewing (1969) noted that the anemia in his Basenjis closely resembled a hemolytic anemia in man that is caused by a deficiency in the red cells of the enzyme pyruvate kinase (PK). Subsequently, Searcy et al. (1971) found that Basenjis with hemolytic anemia were also deficient in PK. Until this point, what slight genetic evidence there was had shown that the condition was recessive, autosomal, and probably caused by a single gene in the homozygous state. The next advance was made by Brown and Teng (1975), who took advantage of the fact that the activity of the PK enzyme can be measured quantitatively to assay PK values in samples of blood from about 1,000 Basenjis. These revealed the useful fact that, among clinically normal Basenjis (which must have included many relatives of anemic ones), there was one group in which units of PK activity ranged from 0.84 to 2.00, and another in which those same units varied only from 0.36 to 0.82. Combining that information with genetic tests indicated that the latter

FIGURE 10-5
The Basenji—not "barkless" as it is commonly said to be, nor is it mute. Weight: about 22–24 lbs. [Courtesy of the American Kennel Club.]

group were probably heterozygotes carrying the gene causing hemolytic anemia, and that the dogs with PK activity above 0.84 units were probably hemozygous for the normal allele of that gene.

Conclusive evidence was provided by Andresen (1977a) from studies of Basenjis in Denmark. His extensive pedigrees of dogs with hemolytic anemia confirmed previous indications that the disease is caused by a recessive, autosomal gene. To that gene he assigned the symbol *pk,* with *PK* for its dominant allele. The pedigrees revealed a number of dogs still living that had been proven by their progeny to be carriers *(PK pk).* With these, he was able to make assays of pyruvate-kinase activity in dogs of known genotype rather than to assume the genotypes from levels of that enzyme in the blood.

In this material (and after special techniques to eliminate white cells), Andresen (1977b) found the mean units of pyruvate-kinase activity in the hemolysates from the red cells to be as follows:

Numbers and Kinds of Dogs	Units
In 13 dogs proven to be *PK pk*	0.60
In these 13, plus 17 more presumed to be *PK pk*	0.63
In 12 dogs presumed to be *PK PK*	1.32

The ranges in pyruvate-kinase activity for heterozygotes and normal homozygotes overlapped slightly, but it was concluded from statis-

TABLE 10-3 Segregation of Hemolytic Anemia in the Basenjis of Brown and Teng (BT) and Andresen (A)

<table>
<tr><th colspan="2" rowspan="2">Parents</th><th rowspan="2">Expected Ratio</th><th colspan="3">Progeny</th></tr>
<tr><th>Normal</th><th>Carriers</th><th>Anemic</th></tr>
<tr><td>BT</td><td>Carrier × *PK PK*[a]</td><td>1 : 1 : 0</td><td>8</td><td>8</td><td>0</td></tr>
<tr><td>BT</td><td>Carrier × *PK pk*[b]</td><td>1 : 2 : 1</td><td>6</td><td>9</td><td>8</td></tr>
<tr><td>A</td><td>Carrier × *PK pk*[b]</td><td>3 : 1</td><td colspan="2">25[d]</td><td>11</td></tr>
<tr><td>A</td><td>Carrier × *pk pk*[c]</td><td>1 : 1</td><td>0</td><td>7[e]</td><td>4</td></tr>
</table>

[a]Homozygous normal.
[b]Carrier.
[c]Anemic.
[d]Includes *PK PK* and *PK pk* in ratio of 1 : 2.
[e]All dogs not anemic must be *PK pk*.
Source: Brown and Teng (1975), Andresen (1977b).

tical analyses that a value of 0.99 units would separate the two classes with errors fewer than 5 percent.

In further studies, Brown and Teng (1975) found two other tests by which carriers could be identified. One of these was the level of adenosine triphosphate (ATP) in erythrocytes of blood that had been stored for 24 hours. It was effective in puppies as well as in adults. The other test was for PEP.[2] It identified carriers when adult, but not as puppies.

The experimental matings of Brown and Teng and the pedigrees compiled by Andresen (1977a) yielded ratios showing conclusively that this anemia in the Basenjis is a simple recessive trait which segregates in clear Mendelian ratios (Table 10-3).

For breeders, the discovery of a test whereby carriers of an undesirable gene can be distinguished from normal homozygotes is a remarkable advance. A laboratory test is far faster and easier than the breeding test which is necessary for detecting most carriers of other undesirable recessive genes. It is possible in this case because neither of the alleles involved is completely dominant or recessive. They are codominant. Each exerts its influence. An easy way to think of such a situation is to remember that it takes two doses of one allele to establish one effect (normality) and two doses of the other allele to produce the opposite effect (anemia). One dose of each results in an intermediate state—clinically normal, but recognizable by special tests.

Among the first 300 samples of Basenji blood tested by Brown and Teng, the proportion considered to be from heterozygotes was 19 percent. Further utilization of these tests and consequent elimination of carriers from reproduction should greatly reduce the incidence of hemolytic anemia in the breed.

[2]Not what you think. In this case, PEP means *phosphoenolpyruvate.*

FIGURE 10-6
A dwarf Malamute showing the out-turned feet, enlarged carpal joints, and bowing of the radius and ulna that are typical clinical symptoms of hemolytic anemia with stomatocytosis. [From Subden et al., 1972.]

Anemic Dwarfs in Alaskan Malamutes

When Subden et al. (1972) made their thorough genetic study of the dwarf Alaskan Malamutes, they called the condition *chondrodysplasia,* but recognized that it differed from other dysplasias of the skeleton. The dwarf dogs were smaller than normal, with enlarged carpal joints, and short front legs bowed inward, so that the general appearance was that of a rachitic animal (Figure 10-6). Both sexes were affected.

From cooperating breeders, records were obtained of litters from ten carrier sires, each of which had been mated with at least two carrier females. These had yielded a ratio of 52 normal : 19 dwarf. Subsequently, in experimental matings, backcrosses of carrier × dwarf produced 11 normal : 8 dwarf. Finally, mating dwarf × dwarf yielded 11 pups, all dwarf. Dwarfism of the type in these Alaskan Malamutes is clearly a simple, recessive, autosomal trait.

No intermediate types were found. Conclusive identification of the dwarf phenotype was possible by 25 days of age with radiological examination of the front legs.

Subsequent studies by Fletch and Pinkerton (1972) and Pinkerton et al. (1974) led to the discovery that all the dwarf dogs had a mild hemolytic anemia. It was associated with a peculiar type of erythrocyte called a *stomatocyte* (translated: "mouth cell"). That name is used because the cell has a well-defined unstained area across its center, which suggests (under the microscope) that the cell has an opening like a mouth.

It is now clear that the dwarfism, chondrodysplasia, and anemia are all expressions of a syndrome induced by one recessive gene in the homozygous state. It seems probable that, as in the gray Collies, the underlying abnormality is in the blood. Just as the lethal gray of Collies led to discovery of the associated cyclic neutropenia, so did the dwarfism of the Malamutes lead to the recognition of what is now called "hereditary stomatocytosis with hemolytic anemia."

Immunogenetics and Polymorphism

Lest any dog owner should get the impression that what has been written thus far in this chapter covers all known genetic variations in canine blood, it is desirable to say a little about subjects included under the banner of the two big words that head this section.

The first one, immunogenetics, refers to hereditary variations in blood groups, antigens, proteins, enzymes, and other constituents that are in the blood cells or on them, in blood serum, or in other body fluids or tissues. We cannot see them, but they can be identified by appropriate tests. We know that they are important, because some of them may have to be considered when blood transfusions are made, and others determine whether or not one body will accept the graft of skin, a heart, or anything else from another body. The second word, polymorphism, merely reminds us (in Greek) that some things can occur in many different forms.

Knowledge of canine immunogenetics has advanced rapidly since the review by Swisher and Young (1961) of what was then known about blood groups of dogs. At the Second International Workshop on Canine Immunogenetics held in Portland, Oregon, in 1974, 46 participants from 18 laboratories in six countries pooled their information and compared ideas to find how much more they had learned about their special interest since the first such workshop was held in Rotterdam two years earlier. Much of what follows here is taken from the extensive joint report of the Second Workshop by Vriesendorp and 25 others (1976).

By no means least of the accomplishments at that meeting was the adoption of standard terminology and symbols for the many variations studied. That was essential if workers in any laboratory were to understand what those in some other had found. It was particularly necessary for the study of genetic variations, because some of the loci in the chromosomes have multiple alleles. Each locus and each allele had to be given a label that meant the same thing in Rijswijk (The Netherlands), in Paris, in Guelph (Canada), and elsewhere.

Variations considered at the Portland conference are summarized (in much condensed form) in Table 10-4. A few comments on that table will help us to understand what it is all about. More important, they may help us to comprehend a little better what we shall be reading in

TABLE 10-4 Some Genetic Variations in Canine Blood

Class	Locus or Loci	Genes or Alleles	Symbol
I. Dog Erythrocyte Antigens, at Locus DEA 1	1	3	DEA 1.1, DEA 1.2
At other loci	6	6	DEA 3 to DEA 8
New, for further study	?	8	Tentatively N 1 to N 8
II. Dog Leukocyte Antigens, Locus A	1	7	DLA-A 1 to A 3, A 7 to A 10
In same chromosome, Locus B	1	5	DLA-B 4 to B 6, B 12 to B 13
Not "assigned"	?	4	DLA-11, R 15 to 17
DLA mixed lymphocyte cultures	1	8	DLA 50 to 57
New, for further study	1	4	V, X, Y, Z
Leukocyte antigens, different locus	1	2	DLB-1
Same, a third locus	1	2	DLC-1
III. Polymorphic Enzymes in Blood Cells			
Phosphoglucomutase-2	1	2	PGM 2-1 and 2
Phosphoglucomutase-3	1	3	PGM 3-1, 2, and 3
Peptidase D	1	3	Pep D-1, 2, and 3
Superoxide dismutase-soluble	1	2	SOD-S 1 and 2
Glucose-6-phosphate dehydrogenase	1	2	G6PD 1 and 3
Glutamate oxaloacetate transaminase soluble	1	2	GOT-S 1 and 2
IV. Proteins in Blood Serum			
Albumins	1	2	Alb A and B
Transferrins	1	4	TfA, B, C, and D
Immunoglobulins	3?	?	IgG, IgM, and IgA
V. Canine Secretory Alloantigens	1?	3	CSA A, Y, and X

Source: Vriesendorp et al. (1976), with the exception of immunoglobulins, which are from Bull (1974).

the future about antigens, enzymes, serum proteins, and other such things in dogs.

Dog Erythrocyte Antigens

The red corpuscles carry *dog erythrocyte antigens* (DEA) which determine at least eight blood groups, and (as those new ones numbered N1–N8 suggest) there are more to be identified in the future. They are called specifically *erythrocyte* antigens to distinguish them from other antigens not carried on the red cells, and they are designated as *dog* antigens just so that they will never be confused with the antigens of man, or of any other species. To investigators in this field, the short label DEA tells exactly what is under study. These red-cell antigens were earlier labelled by letters, then as CEA 1–8, now DEA as shown in Table 10-4 (CEA stands for *canine erythrocyte antigens*).

At the DEA 1 locus, where three alleles are now known (one is called a "blank"), no dog can carry more than two of those three. The same applies to any series of multiple alleles. At each of the other six loci, which are probably in six different chromosomes, there are two alleles, one of which is the mutant that determines a specific antigen and the other its partner allele, which does not determine that same antigen. We can think of the latter as a gene inducing normal blood. Some dogs may carry the mutant gene; others may not. Consequently, any dog can have all of the eight DEA groups, or only some of them.

The frequencies of these DEA antigens were found to vary in mongrels (as one class), Beagles, and Retrievers. The most common one was DEA 6, found in these three classes with frequencies, respectively, of about 74, 76, and 100 percent. For DEA 3, the least common, the corresponding frequencies were only 5, 5, and 0 percent.

Dog Leukocyte Antigens

Dog leukocyte antigens (DLA) comprise what was earlier known as the "major histocompatibility complex." They are responsible for rejection of grafts that are not retained by the host receiving them. At the Portland conference, they commanded more study than anything else. Incompatibility of grafts is an important problem in human medicine. If it can be elucidated by study of the same phenomenon in dogs, it is obviously desirable to learn whatever the dogs can teach us.

Most of the numerous antigens in this group are apparently found at two loci in one chromosome where series of seven and five alleles have been identified. The antigens found in cultures of mixed lymphocytes apparently originate within the DLA complex, but, until it is learned to which locus there they belong, they are labelled as DLA 50 to 57, plus four new ones that have so far qualified only for the obscurity of the letters V, X, Y, and Z.

Polymorphic Enzymes

Man's mutant enzymes have been much studied in recent years. Table 10-4 lists six enzymes (none pathogenic) for which mutant forms in man have also been identified in dogs, and it is certain that the list will be extended.

One not shown in Table 10-4 is the mutation that, by causing subnormal activity of the enzyme pyruvate kinase, is responsible for hereditary hemolytic anemia of the type originally found in Basenjis.

Another reduces activity of the enzyme β-galactosidase, and thus causes accumulation of gangliosides (fatty substances found in ganglion cells of the nervous system) in the brain, spleen, and liver. This leads to degeneration of the brain, and eventually to inability to stand. Two forms of gangliosidosis are known in man, but both were found in one dog by Read et al. (1976). Its pedigree suggested that deficiency of the enzyme is a simple, autosomal, recessive trait.

Levels of catalase, another enzyme in the dog's red blood cells, were found to be determined by a single pair of codominant alleles (Allison et al., 1957). The deficient homozygotes showed no obvious handicap, such as the tendency to progressive oral gangrene which has been reported in humans lacking catalase, but in laboratory tests their blood was found to be particularly subject to hemolysis.

Proteins in Blood Serum

Genetic variations are known in three kinds of serum proteins. Albumins (the most abundant) vary in quantity according to the amount of protein in the diet. Two alleles induce (when homozygous) the AA and BB types, but, as with most other polymorphic variations in the blood, the alleles are codominant and the heterozygotes (AB) can be identified.

Transferrins (Tf) are so called because they carry iron. Four codominant alleles are known, making 10 different genotypes possible. Two of the alleles are rare. When Stevens and Townsley (1970) tested 248 dogs (of nineteen breeds) for transferrins, the TfA type was found in only 8 animals. Five of these were Beagles (among 30 of that breed tested), and 3 were West Highland White Terriers. At the Portland conference, the TfA type was reported only in Beagles and the TfD allele only in mongrels, both types equally rare. The B and C alleles were reported as segregating nicely in three kinds of matings according to the 1 : 1 and 1 : 2 : 1 ratios expected (Table 10-5).

Immunoglobulins play an important role in resistance to infection. According to Bull (1974), three major classes of these have been iden-

TABLE 10-5 Segregation of Transferrin Types in Three Kinds of Matings

Types in Parents	Types in Progeny		
	BB	BC	CC
BB × BC	36	29	0
CC × BC	0	11	10
BC × BC	18	45	16

Source: Vriesendorp et al. (1976).

tified in dogs. One of them (Ig G) has four distinct subclasses. Study of these immunoglobulins at the Portland conference was apparently restricted to those at the Ig M locus, and the two alleles there were designated simply as + (plus) or − (minus). Almost all of 44 Beagles were Ig M+, but in mongrels and Retrievers the frequency of the Ig M− allele was 36 percent.

Canine Secretory Alloantigens

Secreted in the saliva, canine secretory alloantigens are sometimes referred to as digestive tract antigens. Three types of these are recognized, but it is not yet clear how many loci and alleles are responsible for them. The CSA A antigen is associated in inheritance with the DEA 7 antigen, but their exact relationship is not yet clear.

Significance of All This

My chief purpose in reviewing these somewhat esoteric studies in canine genetics is to let dog owners know what is going on. Anyone appalled by the polysyllabic vocabulary of Table 10-3 should relax at once. It does not have to be memorized. Nevertheless, the questions inevitably arise—"How important is it all? What now?"

Blood groups have to be considered whenever transfusions are made in man, but apparently this is not essential in dogs except when repeated transfusions are necessary. This is so, according to De Wit et al. (1967), because nearly all dogs carry the two stronger antigens (now known as DEA 4 and DEA 8), and the others are weak antigens that rarely cause trouble. It is entirely possible that in the future some mutant type may be found that is a troublemaker like the *rh* in man and its counterpart in the horse.

The leukocyte antigens are of importance whenever grafts of any kind have to be made from one animal to another. Even if that does not happen as often in dogs as in man, we should know that dogs vary in histocompatibility, just as do their owners.

Genetic deficiencies in enzymes can be serious. Hemolytic anemia and gangliosidosis give warning that other mutations in enzymes are likely to be recognized in the future, that some will be lethal, and most will at least reduce physiological efficiency. The Swiss zoologist, Hadorn, had a convenient and suitable adjective for such mutations: he called them *subvital*.

Another way in which knowledge of immunogenetic polymorphism may become important to dog breeders is in the determination of parentage. We must remember that with dogs, as in man, there is seldom any difficulty in identifying the mother, and that tests using blood groups do not definitely identify the sire. They do tell what groups the sire could or could not have and thus exclude some suspect males from paternity. Vriesendorp et al. (1973) were able to do this in three cases of disputed paternity in dogs by the use of various antigens, types of serum proteins, and of the enzyme peptidase D. One laboratory in the United States (at Michigan State University, in East Lansing) now makes such tests in dogs, and more may follow. Similar tests are now in use to verify pedigrees in cattle and horses and to identify inbred lines of the domestic fowl. Incidentally, the fact that at the B locus in cattle hundreds of alleles have been identified suggests that what is now known about blood antigens in dogs may be only a beginning.

If the foregoing glimpse of some hitherto unrecognized complexities of a dog's life does not cause us to view with greater appreciation the dog that tips over our garbage can, it may at least let us realize better than ever before how right was Hamlet when he declared:

> "There are more things in heaven and earth, Horatio,
> Than are dreamt of in your philosophy."

Cardiovascular Abnormalities

From studies of congenital heart disease in 35,280 dogs brought during eight years to the clinic of the School of Veterinary Medicine at the University of Pennsylvania, Patterson (1968) showed that

1. The frequency of some kind of congenital heart disease in that large population was 0.68 percent.
2. The common malformations were similar to those in man.
3. Certain defects were significantly more frequent in some breeds than in others.
4. The incidence of such defects in dogs of mixed breeds (which comprised about a third of the population) was only 2.6 per 1,000, a figure significantly lower than the prevalence rate for purebred dogs as a group, which was 8.9 per 1,000.
5. The lowest prevalence rate (per 1,000, after adjustment for age) was 2.9 in the Dachshund, represented by 1,401 dogs. The highest rates were in the Newfoundland (63.5) and the Keeshond (62.5), but the samples of those two breeds were only 64 and 66 respectively. Among 14 other breeds, the prevalence rates varied from 4.0 to 19.9 per 1,000.
6. Six anatomical defects accounted for 72 percent of the 240 cases of congenital heart disease.

The most common abnormality of the heart in dogs is called *patent ductus arteriosus.* In normal development of the fetus, the *ductus arteriosus* is a channel from the pulmonary artery to the heart, but it should become closed before birth. When it fails to do so and remains patent (open), circulation of the blood is abnormal (Figure 10-7). Patterson et al. (1971) found that about half the pups with the duct fully

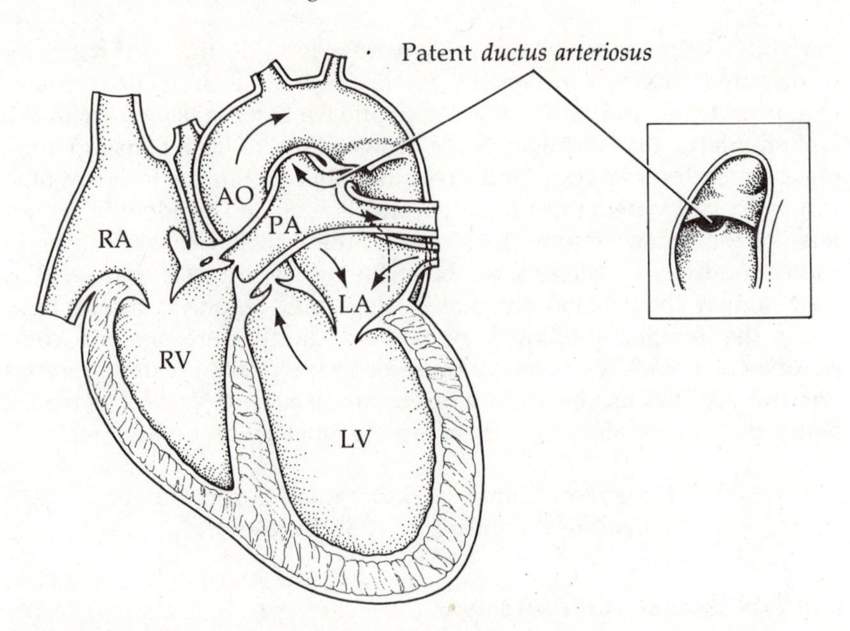

FIGURE 10-7
Effects of patent *ductus arteriosus,* the most common congenital heart defect in dogs. Overloading with blood returned from the aorta through the open duct causes enlargement of the pulmonary arteries (PA), left atrium (LA), left ventricle (LV), and ascending aorta. Arrows show the path of the blood. Enlargement of the left ventricle is extreme. [After Patterson and Pyle, 1971.]

patent developed heart failure and another 15 percent had other abnormalities of circulation. According to Patterson and Pyle (1971), patent *ductus arteriosus* accounts for 25 to 30 percent of all congenital heart disease in dogs.

The facts that breeds differed in the frequencies of cardiovascular defects of any kind and that certain malformations were more common in some breeds than in others indicated that there might be a genetic basis for them. Accordingly, Patterson persuaded breeders to donate for study pups found to have congenital heart disease and, during a period of six years, these dogs were used in experimental matings in which both parents were of the same breed and had the same abnormality. The results, as given by Patterson and Pyle, are summarized in Table 10-6.

It is clear that the five defects[3] there considered are all hereditary. Some idea of that influence is given by the fact that, while the incidence of heart defects (of all kinds) in 2,693 Poodles was found in Patterson's original survey to be only 10.4 per 1,000, when parents of that same breed both having patent *ductus arteriosus* were mated together,

[3]If you want to know anatomical details of these malformations, ask your veterinarian, doctor, or favorite embryologist.

the incidence of that one defect in their progeny rose to the rate of 829 per 1,000.

Similarly, the frequencies of each of the four other malformations were all greatly raised in pups from parents both afflicted with the same specific defect. Only 10 percent of the German Shepherd pups showed the persistent right aortic arch of their parents, but that rate is almost ten times as high as that for congenital heart diseases of all kinds in the 3,077 German Shepherds in the original survey.

Furthermore, the high degree of concordance between pups and parents with respect to each of the five malformations studied shows that there are specific genes inducing specific defects. None of the latter, however, is a simple recessive trait. If it were so, the incidence in pups from parents both affected should have been 100 percent. We can conclude that polygenic inheritance is responsible. Polygenic characters, like these heart defects, which are not manifested at all by some individuals and are expressed in different degrees by others of the same population are sometimes called "threshold" characters. The term implies that in the interaction of genes trying to induce the character and those trying to suppress it (i.e., to maintain normal development) there is a threshold beyond which the former overpower the latter. The degree to which the character is then expressed will vary according to the strengths of the opposing forces. It is simpler to recognize that the trait is hereditary, and that many genes interact to determine whether or not it is manifested, and, if so, in what degree.

In later studies of congenital heart defects in the Keeshond, Patterson et al. (1974) found a range of abnormalities in the conotruncal septum, a partition that (in the embryo) separates the two large arteries that carry blood out from the ventricles. From extensive experimental matings, it was concluded that variations in the degree of abnormality persisting after birth were polygenic in origin. Parents both affected had 152 offspring in 39 litters and 80 percent of these had defects in the septum.

Dog breeders should not think that the congenital heart defects are any reflection on the five breeds listed in Table 10-6. Mutations tending to change the cardiovascular system and circulation of the blood have occurred in vertebrate animals for untold thousands of years. The desirable ones have accumulated and resulted in the four-chambered hearts and efficient blood circulation of mammals. The undesirable ones have been rejected by natural selection. These mutations can occur in any species, and in any breed. In addition to the more common defects listed in Table 10-6, Patterson (1968) found 16 others, most of them comparatively rare. Now that we have been shown so clearly that congenital cardiovascular malformations in dogs are genetic in origin, and even though they are not as common as some other hereditary defects, dog breeders are well advised to avoid breeding from afflicted dogs and their close relatives.

TABLE 10-6 Frequency and Concordance[a] of Cardiovascular Malformations in the Progeny of Dogs with Congenital Heart Disease

Parental Defect	Breed	Matings, Number
Patent *ductus arteriosus*	Poodle	10
Pulmonic stenosis	Beagle	10
Subaortic stenosis	Newfoundland	5
Persistent right aortic arch	German Shepherd	3
Tetralogy of Fallot	Keeshond	4

[a] Same defect in pup as in parents; with partial concordance, the pup had additional cardiovascular malformations, or a defect related to that in the parent.

Source: Patterson and Pyle (1971).

Progeny		Concordance, Percent		
Number	Cardiovascular Malformations, Percent	Complete	Partial	Total
35	82.9	79.3	20.7	100
35	25.7	100	0.0	100
26	38.5	80.0	0.0	80.0
30	10.0	33.3	33.3	66.7
11	90.0	40.0	60.0	100

11 Eyes and Ears

Because dogs and man are both mammals, hereditary abnormalities of the human eye are likely to be found sooner or later in the dog. As some veterinarians are now specializing in canine ophthalmology, we can look forward to a lengthening list of hereditary abnormalities of dogs' eyes to match the much longer one of genetic defects and variations of the eye in man. When Waardenburg (1932) wrote his monograph on that subject, it covered about 100 genetic variations, mostly (except color) pathological in some degree. More recently a treatise by François (1961) required 731 pages and 629 illustrations to elucidate its title: *Heredity in Ophthalmology*. Some day, some specialist in canine ophthalmology will shoot at these marks, but, in the meantime, this author (who is no ophthalmologist) must try to review briefly what he has gleaned from experience and from the literature about hereditary abnormalities of the eye in dogs.

An excellent review of hereditary defects of the dog's eyes by Barnett (1976) came to my attention some months after the manuscript for this book had gone to the publishers. It covers the frequencies of various defects in different breeds, with an opinion on the genetic basis in each case. It also reports that in some kennels cataracts and luxation of the lens have been eliminated by selective breeding, and that progressive retinal atrophy has been similarly greatly reduced in Border Collies.

Progressive Retinal Atrophy

According to Hodgman et al. (1949), progressive retinal atrophy (PRA) became prevalent in Irish Setters in Britain during the late 1920s following considerable pedigree-worship of a famous dog that carried (unknown to the worshippers) a recessive gene causing progressive retinal atrophy. By 1949, the breed association had undertaken a program aimed at elimination of the gene, and their efforts have subsequently met with a gratifying measure of success. Later studies of PRA have shown that the defect is widespread, and, although more common in some breeds than in others, is known to occur in many breeds, some large, some small. Priester (1974) summarized records at 11 veterinary colleges showing its occurrence in 45 breeds.

Age at onset and progress of the disease vary in different breeds and even in different strains of the same breed. In Irish Setters, the first symptoms are sometimes recognizable by eight weeks, but often not until four to six months of age. They include defective vision at night. In the ensuing months (or years), vision becomes progressively worse in dim light and in daylight, until the dog becomes completely blind. Both eyes are affected.

In addition to these clinical symptoms, the clinician notes that the pupil of the eye becomes dilated and the pupillary reflex in response to a bright light is lessened. Constriction of the blood vessels in the retina is evident with the ophthalmoscope. The pathologist notes microscopically degeneration of the rods, cones, and outer nuclear layer of the retina. This begins at the periphery, and grows inward as the disease progresses. In most cases, a cataract is eventually formed in the lens of the eye.

Onset is later in some breeds than in others. In his study of 150 affected Miniature and Toy Poodles, Barnett (1965) found that first symptoms had been noted at four to five years in most cases, but that age at onset varied from seven months to eight years.

CPRA

In extensive studies of retinal atrophy, Barnett (1969b) confirmed previous evidence that in some breeds there is a form of the disease with symptoms different from those just described. Its first symptom is not night blindness, but difficulty in seeing obstacles nearby and even in seeing a person who is standing still. As in the other PRA, vision gradually becomes worse and the dog becomes blind, but the age at onset is later, and some dogs do not become blind until over five years of age. The most conspicuous characteristics of this form of retinal atrophy is seen with the ophthalmoscope. On the tapetum (the layer of the choroid coat that causes dogs' eyes to shine in the dark), there are scattered spots of brown pigment, which are not seen in the PRA described earlier. Degeneration of the retina does not begin at its periphery, but in its central region. For that reason, Barnett called this form *central progressive retinal atrophy*, or CPRA. It was first recognized in black Labrador Retrievers and Rough Collies, but was later found to be widespread in Border Collies (Barnett and Dunn, 1969). It also occurs in other breeds.

Genetic Basis

One may wonder to what extent these two forms of progressive retinal atrophy are really separate diseases, and how much of the differences between them might result from the fact that the expression of a causative gene could vary in different breeds because of its interactions with the many genes that distinguish one breed from another.

For example, variations from that cause might be responsible for the fact that PRA can be detected with an ophthalmoscope in Irish Setters by six months of age, in Norwegian Elkhounds by 2 years, but in Miniature Poodles not until 3 to 4 years of age (Aguirre, 1976). On the other hand, there is no critical evidence to show that PRA is caused by the same mutation in all breeds. The variations just noted could result from different mutations in different breeds.

Barnett (1969b) considered PRA and CPRA as two diseases, the former being a simple, autosomal, recessive character, but believed CPRA to be inherited differently, possibly as a dominant trait with incomplete penetrance.

One difficulty in attempting to assign any genetic basis for PRA and CPRA is that the evidence (such as it is) nearly all comes from the study of pedigrees, and very little of it from experimental matings. In pedigrees, the abnormal animals are always identified, but there is too often inadequate information about the number and sex of normal litter-mates. As a result, the usual criteria for determining the mode of inheritance are missing: There are no Mendelian ratios for F_1, F_2, and backcross generations. Furthermore, pedigrees leave us wondering whether penetrance of the character is complete, or whether in some animals that should show it the genotype is not fully revealed by the phenotype.

It is important to recognize that not every blind dog has either PRA or CPRA. Distemper is known to cause blindness. So do some toxins, other adverse environmental influences, and other diseases. For that reason, a report of a dog in a pedigree as blind is not quite as reliable (for genetic information) as are reports from experimental matings in which all progeny are examined with an ophthalmoscope to determine whether the dog has PRA, CPRA, or neither.

The Electroretinogram

Admittedly, Mendelian ratios are difficult to get with some trait that may not show up until the animals are five years old or more. That problem may now be minimized by the use of a test that reveals PRA at

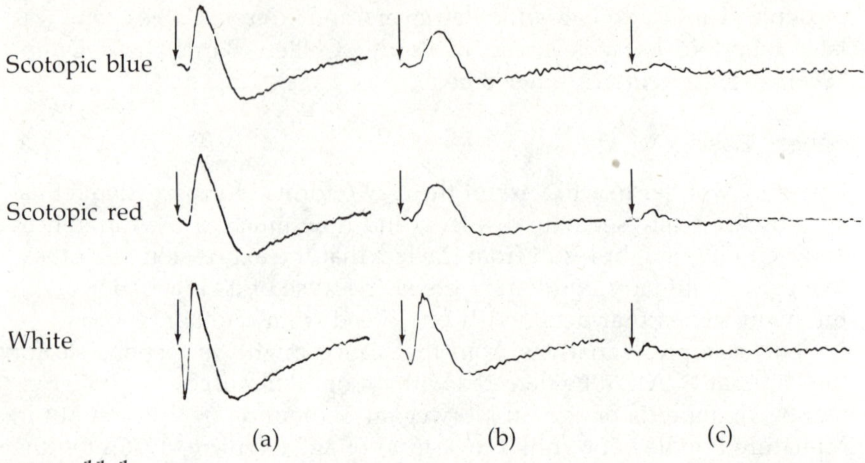

FIGURE 11-1
Electroretinograms of (a) a normal dog, (b) a Miniature Poodle with incipient PRA at 9½ weeks of age, and (c) its mother clinically affected with PRA. Tests with single flashes of three different lights (after 20 minutes in the dark) are shown. Arrows point to the onsets of light flashes. The amplitudes of these and of the subsequent upward and downward waves reveal the differing responses of the three dogs. [From Aguirre and Rubin, 1972.]

a very young age. An electroretinogram provides a record of the response of the retina when stimulated by light. Under anaesthesia the dog is provided with a special kind of contact lens that is connected to a recording apparatus. The resulting electroretinogram records waves of differing amplitudes for dogs that are (1) normal, or (2) affected with PRA, but not yet showing signs that can be detected with an ophthalmoscope, or (3) visibly affected with PRA (Figure 11-1).

With this procedure, Aguirre and Rubin (1972) were able to recognize incipient PRA in Norwegian Elkhounds by 6 weeks of age and in Miniature Poodles by 10 weeks. Because affected Poodles do not usually show the first symptoms of PRA until they are two to four years old, the advantage of the electroretinogram for its early detection is obvious. That advantage might permit genetic test-matings, and would also be desirable for early testing of suspects in kennels where PRA has been known to occur. It is not yet clear whether or not this same procedure is effective in detecting incipient CPRA.

There can be no doubt that both PRA and CPRA, be they one disease or two, are inherited and that breeders should select against them. Reports of one program for doing so are discussed in Chapter 14, on selection.

The Collie Eye Anomaly

The Collies have been deservedly popular for many years, but they are now threatened with the loss of that popularity because of an affliction now known as the Collie eye anomaly. Because that abnormality is widespread in the breed and is commonly supposed to be a genetic defect, it seems desirable to review in some detail what is known about it.

The defect was described under several different names in earlier studies of it, but most investigators now seem agreed on the one just given. The earlier labels, which gave some clue to the nature of the abnormality (for those knowing eyes both inside and out), included *staphyloma,* or *posterior ectasia of the sclera; choroidal changes; anomaly of the optic disc,* or *coloboma;* and *retinal detachment.*

It seems unnecessary to describe here all the varied abnormalities just mentioned. The average dog owner will never see them, although they can be detected with an ophthalmoscope at 6 to 8 weeks of age. All afflicted dogs do not show all expressions of the anomaly. Yakely et al. (1968) recognized seven different kinds of lesions of which the number in any one dog ranged from one to five.

Unlike PRA, the Collie eye anomaly apparently does not lessen ability to see except in dogs that have detached retinas or severe degrees of the other abnormalities. The lesions, unlike those of PRA, do not become progressively worse except that some of them may result eventually in detachment of the retina. Afflicted dogs may be seen to tilt the head to one side as they adjust to any difficulties of vision. In the absence of conspicuous clinical symptoms, the safest plan for the breeder is to have suspect animals examined by a veterinarian who has an ophthalmoscope.

If figures given by various authorities on the prevalence of this disease among Collies are correct, anyone taking a Collie pup for ophthalmoscopic examination had better be prepared for bad news. At a symposium on the condition sponsored by the American Veterinary Medical Association in 1968, Donovan et al. (1969) reported its frequency among some 7,000 Collies examined by them to be 90 percent. Corresponding figures from other sources are only slightly lower. As with similar data for other diseases, one always wonders whether the veterinary clinic got only the diseased dogs, while the healthy ones were kept at home. With the Collie eye anomaly, however, symptoms are absent or barely perceptible; hence it seems probable that Donovan's large sample of the breed was a reliable one, and that, in the years before 1968, the chance of a Collie's not having the abnormality was only about 1 in 10. That chance should be now be much better because of over a decade of selective breeding against the defect.

Genetics—and Second Thoughts

In well-planned matings arranged by Yakely et al. (1968) to elucidate the genetic basis for the Collie eye anomaly, evidence was obtained which suggested that the condition is a simple recessive trait. The original P_1 cross of affected × normal was an outcross of affected Collie females to a male of a breed in which the disease was not known to occur—a Doberman Pinscher. To make doubly sure that the male was contributing no visible eye defects to the experiment, he was carefully examined and found to be sound.

In the F_1 generation, all 20 animals had normal eyes. In six subsequent backcross litters and two of an F_2 generation, the ratios of normal : affected were 20 : 13 and 13 : 2, respectively, both within limits for the expectations of 1 : 1 and 3 : 1 for segregation of a simple recessive character, but both a little short on affected dogs. Similar results were reported for extensive matings within the Collie breed by Donovan and Macpherson (1968). Both teams of investigators found the defect to be equally common in males and females, and, as a result of their studies, it is now generally stated in the veterinary literature that the Collie eye anomaly is a simple, autosomal, recessive trait.

Because some of the lesions in eyes that show the anomaly resemble some of those found in dogs homozygous for the merle gene, it was thought by Roberts (1967) that the two conditions are related, and that the anomaly is associated with dilution of color in the coat and eye. Critical evidence to support that view has not yet been adduced.

And now for the second thoughts. In spite of the evidence just cited, I do not believe that the Collie eye anomaly is any simple Mendelian recessive trait.[1] Some of the reasons for that heresy are as follows:

1. The ratios reported for backcross and F_2 generations would be equally valid if inheritance were polygenic.
2. It is difficult to believe that Collie breeders could raise (by selection) the frequency of a recessive trait that they can seldom detect to such a level that 90 percent of the breed were found to have it.
3. The range in expression of the defect is typical of polygenic inheritance.
4. The genetic evidence said to show that the anomaly is a simple recessive trait does not stand up under critical examination.

Reexamination of the Evidence

Misinterpretation of the data has resulted in at least one case from lumping together in one class (a backcross or an F_2 population) records that completely obscure extreme differences among sires in the proportions of their progeny that had the anomaly. Fortunately, Donovan and Macpherson listed separately for each litter not only the sire and dam, but also the numbers of normal and affected progeny. Three sires, from which the progeny belied the assumption that the sire was heterozygous for a gene causing the defect, are listed in Table 11-1. Of the probabilities listed therein for the sires, all but one (0.083 for ♂ B15) tell us

[1]In all fairness, readers should be told that in some field of genetics I have a well-earned reputation as an incorrigible disbeliever.

TABLE 11-1 Progeny-tests of Three Sires Supposed to Be Heterozygous for a Single Gene Causing the Collie Eye Anomaly

Population	Sire	Litters	Progeny				Probability of Deviation, P^a
			Observed		Expected		
			Normal	Affected	Normal	Affected	
Backcross	B 15	3	21	4	12.5	12.5	.0004
Backcross	B 21	3	5	15	10	10	.0148
	Both		26	19	22.5	22.5	.0693
F_2	B 15	2	13	1	10.5	3.5	.0832
F_2	C 101	4	11	17	21	7	.0001
	Both		24	18	31.5	10.5	.0052

[a]For explanation of the values for *P*, see text.

Source: P values from Warwick (1932); original data of Donovan and Macpherson (1968).

that the ratios in the progeny deviate so much from those expected that we must reject utterly any notion that a single gene is involved. Even if B15's F_2 ratio of 13 : 1 could be expected by chance about eight times in 100 similar trials, that probability of 0.0004 (4 chances in 10,000) for his backcross family tells us that he was not heterozygous for a single causative gene.

It should be noted that the combined data in backcrosses for the two sires make a ratio of 26 : 19, not too far from a 1 : 1 ratio to happen by chance about 7 times in 100 similar trials. That happened because an excess of affected progeny from one sire was counteracted (in part) by an excess of normal progeny from the other.

From their matings of affected × affected, Donovan and Macpherson reported 103 progeny, all affected except one. Such matings provide, in most cases, the critical proof that some trait under study is really a simple recessive character. That test, however, applies to mutations that are uncommon. It does not apply at all to conditions such as the Collie eye anomaly, which has been said to occur in 90 percent of the breed. With any such frequency, one might well say that the anomaly is the normal condition and that unaffected eyes are the exceptions! It is not surprising that from any matings of Collies both affected the frequency of the anomaly in their progeny should rise from the normal 90 percent to 99 percent. Moreover, half of the 24 litters contained only one or two offspring, a total of 19. With larger litters, there might have been a few more normal offspring than the single one reported.

A Suggestion About the Anomaly

In the records of the symposium mentioned earlier, it is possible that, from the standpoint of breeding Collies with normal eyes, the most valuable suggestion was made by Rubin (1969). To him, the extremely high frequency of the anomaly suggested that the breeders might be selecting for some characteristic "which dictates that the ocular anomaly has a high chance for accompanying it."

A linkage of genes on any one of the dog's 39 pairs of chromosomes would provide exactly that high chance. It seems more likely, however, that some inherited trait considered an essential breed characteristic might be involved. It could even be some characteristic considered indispensable in the show-ring! From there, it could easily move into the best social circles (i.e., the pedigrees of champions) and thus travel afar.

The Collie eye anomaly occurs in Collies, and in the much smaller Shetland Sheepdogs, but, according to Barnett (1969a), not in Border Collies. It is apparently unknown or rare in other breeds.

Skulls of both the Collies and the Shelties are extremely dolichocephalic (long and narrow); those of Border Collies and most other breeds are less so. One is tempted to speculate that the narrow heads of the breeds afflicted with the Collie eye anomaly might have something to do with their problem. It is easy to think so because the three lesions most common in that disease—abnormally tortuous blood vessels in the retina, dysplasia of the choroid and retinal layers, and excavation of the optic disc—all suggest that they may be caused by unusual pressure on the eyeball from behind and outside it. Could those beloved narrow heads be distorting the size or shape of the orbits beyond the limits of biological efficiency? Here is a nice problem for some biologist to explore.

In their extensive studies of the shape of the head and skull in different breeds of dogs, Stockard and Johnson (1941) found that, when breeds differing greatly in shape of the head were crossed (e.g., French Bulldog × Dachshund), the F_1 progeny were very uniform in that respect. In dogs of the F_2 generation, however, variability in shape of the head was conspicuous. That is exactly what was to be expected if size and shape of the head are determined by many genes. Undoubtedly each parental breed contributed its quota of genes.

In the matings of Yakely et al. (1968) involving the Collie eye anomaly, the F_1 generation from the cross of normal × affected contained 20 animals—all normal. This was to be expected if the two conditions depended on a single pair of alleles, with dominance of normal and recessivity of the anomaly. That was the interpretation of the investigators.

It seems quite probable, however, that the Doberman Pinscher sire contributed to the experiment not only his good, normal eyes, but also enough genes to widen the skulls of his F_1 offspring beyond the range of the genes for narrow skull that they got from their Collie mothers. No wonder that all 20 had normal eyes!

Breeding Against the Anomaly

Let no Collie breeder fear that to get normal eyes he must outcross to Doberman Pinschers or to any other breed. The useful data of Donovan and Macpherson (1968) show that, within the Collie breed, dogs differ greatly in the degree of susceptibility to the anomaly that is transmitted to their offspring. In Table 11-1, differences in that respect between Sire B 15 and Sire C 101 are conspicuous:

	Litters	Progeny	Affected
Sire B 15	5	39	12.8 %
Sire C 101	4	28	60.7%

With smaller numbers (as in single litters), differences such as these in progeny tests are less likely to be noticed, but they do occur. One of the females mated to ♂ C 101 had a litter of eight in which she totally defied the theory that a single gene was involved by substituting for the expected ratio of 6 : 2 one of 0 : 8. Every pup was afflicted with the anomaly. The probability of that happening by chance alone, with a simple recessive trait, is less than 0.0004!

By watching for differences like these, with routine ophthalmoscopic examinations, and by selecting dogs for breeding according to their progeny tests, it should not be too difficult to reduce greatly the frequency of the Collie eye anomaly. The subject is considered further in the chapter on selection.

Cataracts

Cataracts have been reported in at least eight breeds and four countries, so there would seem to be no reason why they should not be expected in any breed, anywhere. They do not permit easy study by the pathologist or the geneticist, because cataracts differ not only in size, shape, and site of origin, but also in the ages at which they can be recognized. Differences reported among breeds in the incidence of cataracts may therefore depend to a considerable extent on variations in the age at diagnosis, on methods used therefor, and on differing opinions among investigators on the extent to which tiny opacities in the lens should be counted as cataracts.

Among American Cocker Spaniels of unstated ages examined by Yakely et al. (1971), about 88 percent had opacities in the lens, but these varied in the degree of severity. In 11 percent of the 285 dogs examined, the cataracts had developed far enough to cause blindness. In 1,974 dogs of other (or mixed) breeds, the proportion showing cataracts was 37 percent, less than half that in the Cockers. Heywood (1971) found the proportion of 1,064 Beagles showing developing cataracts to be only 3.6 percent up to one year of age, but one wonders what the score might have been in those same Beagles two years later. Presumably there are different types of cataracts and differences among breeds in age at onset. In Pointers in Norway, Høst and Sreinson (1936) studied a type of hereditary cataracts that became evident only at two to three years of age, but developed rapidly and made the dogs blind.

While most reports agree that cataracts are hereditary, attempts to discover any genetic basis have not yielded conclusive results. The defect was said to be a simple, recessive, autosomal trait in Standard Poodles by Rubin and Flowers (1972), in Old English Sheepdogs by Koch (1972), and in Miniature Schnauzers by Rubin et al. (1969). Perhaps it is, but in the first two of these reports, not all dogs in every litter were examined, and in the somewhat limited data for the Miniature Schnauzers, the matings of normal × normal yielded a slightly higher proportion of pups with cataracts than did matings of normal × cataracts.

Among 1,129 Beagles, Andersen and Schultz (1958) found only 1 with cataracts, a male. When he was mated with 5 other Beagles (all free of cataracts), the ratio in his progeny, as determined by postmortem examination at 41–60 days of age, was 23 with cataracts to 2 without! If this was dominance, that ratio was beyond the realm of possibility for any single gene, and the investigators wisely concluded that speculation about the mode of inheritance would be inadvisable without further data.

The difficulty in making conclusive genetic studies of cataracts in dogs is well illustrated by the experience of Olesen et al. (1974) with Cocker Spaniels. A breeder asked for help because about 10 percent of puppies ready to sell at over four weeks had white spots in their eyes. These did not occur in any of the Miniature Poodles and Dachshunds in the same kennel. Furthermore, all of the puppies with cataracts were sired either by one male or his son. When these two were excluded from breeding, no more cases occurred. Four test-matings of affected × affected were made, and three of these yielded only normal pups, none with cataracts. From the fourth mating, there were eight with cataracts and two without. Clearly, the defect was inherited. Equally clearly, the mode of inheritance was complex.

From a cross between female Cocker Spaniels with cataracts and a Norwegian Elkhound that was free of ocular defects at nine years of age, Yakely (1975) obtained an F_1 generation in which none of the 25 dogs examined had cataracts. When 5 of the F_1 males were backcrossed to Cocker Spaniels having cataracts, 34 offspring were examined, and among these only 6 showed cataracts.

It seems clear that cataracts can be genetic in origin, that some breeds have more than others, that some kinds of cataracts are evident early in life, others later, and that affected dogs and their close relatives should not be used for breeding.

TABLE 11-2 Segregation of Hemeralopia in Matings of Alaskan Malamutes

Parents	Litters	Progeny	
		Normal	Hemeralopic
Both normal	1	11	0
Carrier ♂ × normal ♀	2	13	0
Hemeralopic ♂ × carrier ♀	5	17	15
Both carrier	3	17	8
Both hemeralopic	6	0	25

Source: Data of Rubin et al. (1967).

Other Genetic Abnormalities of the Eye

Hemeralopia

The inability to see in bright light, without loss of vision in dim light, hemeralopia, was discovered in Alaskan Malamutes and Poodles by Rubin et al. (1967). The peculiarity was determined in bright sunlight by observing the behavior of the dogs, and in dim light by "scotopic flicker fusion electroretinography." By eight weeks of age, affected dogs would bump into obstacles in bright light, but avoid them in dim light. With the flickering device, the hemeralopic dogs could be identified at six weeks.

From seventeen matings of the Alaskan Malamutes made to elucidate the mode of inheritance of this peculiarity, 110 pups were born, of which 106 survived to be tested. As the character was autosomal, numbers in the two sexes could be combined. These showed clearly that hemeralopia is a simple recessive character (Table 11-2).

Retinal Dysplasia

Bilateral detachment of the retina observed in related Bedlington Terriers by Rubin (1963, 1968) led him to make a series of experimental matings to determine any possible genetic basis for the defect.

An outcross to a mongrel yielded only normal pups in the F_1 generation, but, when some of these were backcrossed to an afflicted dog, the resultant ratio of 14 normal : 9 afflicted indicated that a single recessive gene was involved. So did an F_2 ratio of 7 : 2, and the fact that in two litters from parents both afflicted all 10 progeny had retinal dysplasia. With admirable caution,[2] Rubin (1968) concluded, "Although it cannot be stated with absolute certainty that retinal dysplasia is determined by a single allele, this is the most likely explanation."

Retinal dysplasia was also observed in inbred Sealyham Terriers by Ashton et al. (1968), whose limited evidence suggested that it is a sim-

[2]Other veterinarians, please note.

ple recessive defect. That was also the case in Labrador Retrievers studied by Barnett et al. (1970) in Sweden and in Britain. It occurred only in certain strains, but all cases in both countries could be traced back to one dog born in 1934. In fifteen litters from clinically normal parents in Sweden, there were 99 normal : 26 affected. In three litters from parents both affected, mortality in the 17 pups was so high that only 8 lived long enough for diagnosis. All were affected.

Abnormalities of the Lids

Most lists of hereditary abnormalities of the dog's eyes include entropion (eyelid turned in), ectropion (eyelid turned out) and trichiasis (facial hair growing toward the eye), and distichiasis (extra hairs growing from the lid margin), but genetic evidence seems limited to reports that these conditions are more common in some breeds than in others. The extensive survey by Hodgman (1963) turned up 485 reports of entropion in sixty breeds. The fact that there were 91 cases in Cocker Spaniels could mean either (1) that the affliction is more common in that breed than in others, or (2) that the veterinarians contributing to the survey saw more Cocker Spaniels than dogs of other breeds. In Australia, Johnston and Cox (1970) ranked entropion as fifth in number among 1,371 cases of abnormalities reported in dogs. Of their 145 dogs with entropion, more than half were Labrador Retrievers.

Breeders should not use dogs afflicted with any of the conditions just listed and should also avoid their close relatives. Noting that in his experience breeds with much loose skin on the head (St. Bernard, Clumber Spaniel, Bloodhound, and Chow Chow) seem to have much more trouble with eyelids than others, Crawford (personal communication, 1977) suggests that some selection against loose skin (and hence for tighter eyelids) might reduce such conditions.

Luxation of the Lens

Most lists include luxation of the lens, and with little more genetic evidence than its high incidence in Fox Terriers and Sealyham Terriers. Hodgman's survey (1963) turned up 152 cases in thirty-one breeds (mostly in nine), with Fox Terriers heading the list, and making up almost half of the total number. There were only 12 cases in Smooth-haired Fox Terriers, but 60 in Wire-haired Fox Terriers.

Persistent Pupillary Membrane

Some investigators believe persistent pupillary membrane to be hereditary because it is apparently more common in Basenjis than in other breeds. Normally, the membrane covers the pupil during fetal growth, but it degenerates in most dogs just before birth and disappears by three to five weeks. In Basenjis, it apparently persists several weeks longer. Roberts and Bistner (1968), who observed the condition in 200 Basenjis, and Barnett and Knight (1969), who examined 222 dogs of that same breed, all believed it to be genetic in origin. Roberts and Bistner stated that the condition had no apparent serious effect on vision. However, opacities of the cornea and lens may be caused by persistent pupillary membranes (Riis, personal communication, 1977).

Aphakia

Congenital absence of the lens, or aphakia, is a rare defect. It was reported in five St. Bernards by Martin and Leipold (1974). All had impaired vision since birth. As at least four of these were said to have occurred in different litters sired by one dog, the abnormality may be a hereditary defect. Further evidence is desirable.

Homozygous Merles

The eye defects of dogs homozygous for the merle gene were mentioned in Chapter 2. They are pleiotropic effects of the gene that causes dilution of color, and vary all the way from complete absence of one or both eyes and differing degrees of microphthalmia to eyes of full size that are pale blue and more or less sightless.

The associated anatomical abnormalities within the eyeball have been fully described by ophthalmologists. As the merle gene is only incompletely dominant, the heterozygous merles frequently show some degree of abnormality in the eye. In these, it is unilateral more often than bilateral.

Ears

In contrast to all the genetic abnormalities in dogs' eyes, there seems to be only one serious defect in the ears, and that is deafness. Fortunately, it is not common. In some species, white color in the coat is associated with deafness, as, for example, in blue-eyed white cats. A comparable deafness is usual in Collies homozygous for the merle gene. These have pale blue eyes, and are almost always completely deaf. Other color varieties proven to carry the merle gene are the dappled Dachshund, merle Shetland Sheepdog, and Norwegian Dunkerhund. It is possible that deafness could be associated with pale eyes in any of these breeds. That applies also to Foxhounds, Old English Sheepdogs, and Harlequin Great Danes, all of which have been suspected to carry the merle gene, but are not yet proven to do so.

Deafness in Dalmatians and Other Breeds

A different kind of deafness, not associated with white coat color, has been reported in many different breeds. There is a belief among dog breeders that it is more common in Dalmatians than in some other breeds. Hodgman's survey of veterinary clinics in Britain turned up more reports of deafness in Dalmatians than in any other breed, but 22 breeds had one or more cases.

A good attempt to find a genetic basis for deafness in Dalmatians was made by Anderson et al. (1968) in Sweden. Starting with a pair of dogs that had been discarded by a breeder for deafness, and adding a similar pair later, they measured ability to hear in 49 descendants born in twelve litters. This was done under conditions that were uniform for all dogs tested. To ensure that uniformity, every dog was firmly shackled by the neck in a stand specially designed for the tests, so that the heads were in every case at a constant distance (40 centimeters) from the loud-speaker through which sounds for the test were transmitted. The dogs were tested after four months of age, and all were tested at least twice to ensure reliable determinations. Standard procedures of audiometry were used. Anatomical abnormalities in the inner ears of deaf dogs were studied in detail.

It was found that in each of the two pairs of dogs with which the experiment started one member was totally deaf and the other had only remnants of hearing. In their 49 progeny, only 3 were completely deaf, 5 had remnants of hearing and in three more ability to hear was slightly reduced. The other 38 were normal. In the 15 dogs with any degree of deafness (including the two parental pairs), there were 10 ♂♂: 5♀♀.

The experiment provided no evidence that deafness was caused by any single dominant or recessive gene with complete penetrance. Anderson et al. thought that a sex-linked gene with varying expressivity might be involved, but evidence on that point is far from conclusive. More significant as a warning to breeders is the fact that, among the 49 descendants of deaf dogs, 22 percent were similarly afflicted.

2 Other Hereditary Abnormalities

Most of the genetic disorders that make problems for dog breeders have been reviewed in previous chapters. Others not common enough to constitute a menace in any breed, or for many breeders, are briefly considered in this one. Any of them could cause serious problems for the breeder in whose kennel they occur.

Hernias

Umbilical Hernia

Umbilical hernia, in which a bit of intestine or omentum protrudes through the abdominal wall without any rupture of the skin, can be attributed to failure of the umbilical ring to close. The condition is found in several species of mammals, and dogs are not exempt.

In a review of the cases reported in several species by a number of veterinary hospitals, Hayes (1974) had 382 cases in dogs, with forty-five different breeds represented in the group. His analysis indicated somewhat higher risk of umbilical hernia in the Airedale Terrier, Basenji, Pekingese, Pointer, and Weimaraner. It must be remembered, however, that these calculations were based on the numbers of dogs in the hospitals, and that, while such records for several hospitals can yield large numbers, one may well question whether or not they provide adequate samples of the breeds.

Umbilical hernia in Cocker Spaniels was studied by Phillips and Felton (1939). It appeared within five weeks of birth, was usually small, and sometimes disappeared soon after weaning. Those hernias that persisted in adults became proportionately smaller than in the puppies. In one kennel, parents both normal yielded offspring (nine litters) among which about a third showed hernia, but, among 10 puppies from parents both herniated, 6 showed the trait. Phillips and Felton concluded that susceptibility to the condition is genetic, recessive, polygenic, and independent of sex and color.

Experimental investigations to demonstrate a genetic basis for umbilical hernia in dogs are not necessary. Enough evidence has been provided from studies with cattle and rats to prove that the underlying weakness is hereditary in those species. It is likely to be the same in other mammals. Moore and Schaible (1936) selected for hernia in their rats and raised its frequency from an initial 2.7 percent to 71.2 percent. Moreover, they ended up with hernias far bigger and better than the little ones with which they had started selection.

The message for dog breeders seems clear. Don't breed from any animals with umbilical hernia—or from one in which the defect has been surgically "corrected." Eliminate it by progeny-testing, as with any other polygenic trait.

Inguinal Hernia

With inguinal hernia, loops of the intestine descend through the inguinal canal into the scrotum. The survey by Hayes turned up 88 cases in several breeds. Among those with a comparatively high incidence was the Basenji, which Fox (1963) had previously found susceptible to the defect. Others with risk above average were the Basset Hound, Cairn Terrier, Pekingese, and West Highland White Terrier. Crossbreds had a relatively low risk. By selection for inguinal hernia in swine, Warwick (1931) raised its frequency from 1.7 percent in the original unselected stock to 90 percent in only six generations. Further warning to dog breeders seems unnecessary.

The Digestive System

We have dealt in earlier chapters with two hereditary abnormalities that interfere with efficient performance of the digestive system. One is the glossopharyngeal defect, or bird-tongue (Chapter 1), which prevents newborn puppies from swallowing; the other is cleft palate (Chapter 7), which causes suckling pups to lose the milk through the nose. There are two other defects that should be mentioned, even though we know very little about them.

Gingival Hypertrophy

Burstone et al. (1952) reported four male Boxers that had hypertrophied gums (gingival hypertrophy). Histological examination of the lesions revealed some cystic changes and fragments of bone in the fibrous underlying tissues. These cases merit mention in this book because three of the dogs were grandsons of a common grandsire, and the fourth was a great grandson. To the best of my knowledge, no other such cases have been described in the 25 years since those four were reported.

Another abnormality of the gums is the hereditary hyperplastic gingivitis that Dyrendahl and Henricson (1960) found to be hereditary in foxes. (Readers who now stop reading on the ground that foxes are not dogs should remember that both species are true Carnivores; that their teeth, jaws, and gums are almost identical; and that mutations in either species could occur in the other). In the foxes, swelling of the gums around the posterior cheek teeth was evident as early as six months after birth. A tumorous mass gradually spread forward, and eventually both upper and lower jaws were completely involved. At two to three years, only the highest crowns of the teeth were visible, and some animals were unable to close their mouths.

Among 465 progeny from matings of parents both affected, the number showing the gingivitis was 143. The investigators concluded that the defect is caused by an autosomal recessive gene with penetrance of about 30 percent.

Dilation of the Esophagus

This condition has often been labelled esophageal achalasia in the veterinary literature, but it is now recognized that use of the term *achalasia* is a misnomer. The defect causes regurgitation of food and is easily recognized, after pups go on solid food, by their habit of vomiting several times in one day. It was believed by some investigators that the regurgitation resulted from failure of a sphincter muscle at the lower end of the esophagus to relax and thus to permit passage of food into the stomach. Such a failure of any sphincter muscle could properly be considered achalasia. It is now recognized, however, that the dog does not have a sphincter at the lower end of its esophagus, and that regurgitation of food results from a contraction of the gullet. Contraction toward the lower end of the gullet and resultant congestion above it cause dilation of its upper region.

Various ways of coping with this abnormality are reviewed by Clifford et al. (1972). The methods range from small feedings in an elevated container through assorted drugs to surgery. Apparently some dogs "survive and thrive" without treatment of any kind.

Most writings on dilation of esophagus state that it is hereditary, but none that I have read give any conclusive evidence on that score. Presumably the assumption of inheritance is based on records of several cases in related animals, or in some one breed. On the general principle that where there's smoke there's fire, one cannot ignore reports of familial incidence, but neither can they be accepted as conclusive evidence.

Osborne et al. (1967) cite reports of this condition in 14 breeds. Their own extensive pedigrees of six affected Wire-haired Fox Terriers that came from two kennels but were traced back several generations to a common ancestor are not really evidence of a genetic basis for dilation of the esophagus. In most pedigrees of purebred dogs of any one breed, there can be found, if one goes back far enough, names of the glorious champions of yesteryear. They are ancestors common to many dogs of the latest generations whether these descendents vomit their food or not.

It is to be hoped that some day, somewhere, someone will arrange experimental matings (of the kind shown in several tables of this book) that will reveal whether or not dilation of the esophagus is hereditary, and, if so, how. This may not be easy, because it is not yet clear what proportion of affected dogs survive to breeding age. In any case, it would help to know the proportions affected in different sire-families, in several litters from one dam, and in several litters from parents both of which have produced pups that had the dilation. If it is hereditary, it need not be a simple dominant trait, or a simple recessive; it could be polygenic. And, as we learned from the Swedish foxes described in the preceding section, a genetic defect can be hereditary and recessive, but, because of incomplete penetrance, still show up in only about 30 percent of the progeny (instead of 100 percent) from parents both affected.

The Urogenital System

One abnormality in the urogenital system that has long been considered hereditary has already been discussed in Chapter 4, where it merited a few paragraphs as a possible example of sex-limited inheritance. It is cryptorchidism. There can be no doubt that it is sex-limited, but the extent to which it is genetic in origin remains to be determined by evidence more critical than any thus far adduced. We can now pass on to some hereditary variations in the urinary system.

High Uric Acid in Dalmatians

In most vertebrates, substances in the food called purines are broken down during digestion to uric acid. That is then converted by an enzyme, uricase, to allantoin and excreted in the urine. Exceptions include man, the chimpanzee—and Dalmatian dogs. In all of these the chief end-product of the metabolism of purines is uric acid.

This distinction of the Dalmatians from other dogs has been explored by many investigators. It is now clear that the Dalmatians do

FIGURE 12-1
The Dalmatian—justifiably proud of his distinctive spotting, but completely unaware of his unique physiological distinction. Ch. Battered Bentley in the Valley. [Courtesy of Mrs. Lois Meistrell, Pownal, Vermont.]

not lack uricase, but that uric acid is not resorbed by the kidney tubules in Dalmatians as it is in other dogs. According to Harvey and Christensen (1964), the underlying abnormality may be in the process by which uric acid is transported within the kidney.

This difference between Dalmatians and other dogs is so great that in their genetic studies of it Trimble and Keeler (1938) were able to sort their dogs into two classes that did not overlap. The amount of uric acid excreted in 24 hours per kg of body weight was 4 to 10 mg in those with low uric acid, but over 28 mg in those with high uric acid. They showed that the difference between the two kinds is determined by a single pair of autosomal alleles, that high excretion is recessive, and that it is entirely independent of the black spotting which is a breed character in Dalmatians (Figure 12-1).

Because some kidney and bladder stones have a high content of urates, one might expect to find such stones more common in Dalmatians than in other breeds. Seven of nine unrelated Dalmatians examined by Keeler (1940) were found to have kidney stones. White et al. (1961) analyzed 127 urinary calculi from 26 breeds of dogs, and found that they were mostly phosphate, oxalate, or cystine in various breeds, but those from Dalmatians were mostly urates. The risk of developing such stones can apparently be lessened somewhat by the use of certain drugs.

For the philosophers among us, the Dalmatians pose an interesting case. How did they get their remarkable physiological distinction by which they differ from other breeds? They originated in the province of Dalmatia on the west coast of what is now Yugoslavia. They were bred to run behind carriages or coaches—not to excrete uric acid in abnormal amounts. Although the founders of the breed could select for the distinctive black spotting which is an indispensable breed character of the Dalmatian, they could not select for high uric acid. They did not even know it was there. The most likely explanation is that the mutation causing it occurred in one of the original dogs from which the breed arose.

In any case, the Dalmatians merit a special rub behind the ears because they prevent us from making arbitrary definitions of normality. When what is normal for the breed is abnormal for the species, who is to say what is normal and what is not?

Cystinuria

Some dogs excrete in the urine abnormally high levels of the amino acid, cystine. They are more subject than other dogs to the formation of urinary calculi composed mostly of cystine. According to Bovee and Segal (1971), this peculiarity of metabolism occurs in many breeds, but is apparently most common in Dachshunds and is responsible for about 10 percent of all calculi in dogs. It is said to be hereditary in the dog (as it is in man) but the genetic evidence consists only of its occurrence in animals that can be traced to common ancestors. It is said to occur only in males.

Renal Cortical Hypoplasia

The three words *renal cortical hypoplasia* describe a pathological condition in the outer layers of the kidney. During eleven years, Krook (1957) examined 40 cases of it in Stockholm. Nearly all were in Cocker Spaniels. At that period, the number of Cocker Spaniels in Stockholm was over 2,100, so the disease is evidently rare. Males and females were affected in about equal numbers. Both kidneys were affected, and the disease was lethal by about 12 months of age. It seems probable, however, from other studies, that Krook got only the most severe cases, and that milder degrees of the same disease might not be detected without analysis of the urine. He considered it hereditary.

A disease of the kidneys in Cocker Spaniels, apparently identical with that described by Krook, was studied in Switzerland by Freudiger (1965). He recognized three different grades of severity. Age at onset varied from three months to two years. Dogs most severely affected died (from uraemia) before two years of age, but those with milder grades of the disease did not die during more than two years of observation. Clinical symptoms in the severe cases included excessive urination, intense thirst, vomiting, diarrhea, and dehydration. These were less evident or completely lacking in milder cases, some of which could be recognized only by higher levels of urea in the urine.

In Freudiger's experience, this kidney disease was confined to two kennels of particolored Cocker Spaniels, and not seen at all in dogs with solid color. Over a period of about ten years, the incidence of renal cortical hypoplasia (including mild cases) was approximately 16 percent in one kennel and 20 percent in the other. Of 26 cases, 10 were severe, 9 of Grade II, and 7 mild. One female classified as Grade III only (mild) had a litter of 5 pups, among which 4 had the disease in its most severe form.

There can be little doubt that the condition is hereditary, but the exact genetic basis is not clear. The variations in age at onset and severity suggest that it may be polygenic, but, since mild cases may be undetected, further studies based on levels of urea in the urine are desirable.

Other Disorders

It is not clear whether the condition described by Finco et al. (1970) as familial in Norwegian Elkhounds is the same as that just described. Three of their six diseased dogs resulted from mating their sire back to his dam. All three showed the disease before four months of age.

In related Beagles, Fox (1964) found five pups that had only one kidney. Two that died had interstitial nephritis, and two that were clinically normal showed cystic lesions in the cortex and medulla when examined postmortem. Absence of one kidney in dogs has been reported by others, usually under the name of renal agenesis.

Another rare defect that may have some genetic basis is called *ectopic ureters.* In affected dogs, the ureters empty into the vagina instead of the bladder, and the result is incontinence. Seidenberg and Knecht (1971) had 10 cases in eight years, all females. In another 11 cases reported in the literature, all but one were females. It was suggested that males might have a corresponding abnormality but show no clinical signs because their external urethral sphincter muscles prevent incontinence. Petty (personal communication, 1975) reported 3 cases of ectopic ureters (all females) in two litters of Newfoundlands from one sire. Further evidence is desirable.

Endocrine Secretions

Diabetes

Like man, dogs can develop diabetes from failure of their islets of Langerhans (in the pancreas) to produce enough insulin for adequate metabolism of sugars in the diet. The condition is hereditary in man, and is presumed to be the same in dogs; but the genetic basis is unknown. Wilkinson (1960) noted that, among 56 diabetic dogs examined, almost a third were Dachshunds. Two other hereditary malfunctions of the endocrine glands in dogs are described briefly as follows.

Pituitary Dwarfism

The anterior lobe of the pituitary gland (also termed the adenohypophysis) secretes a hormone that is necessary for normal growth. A hereditary inability to produce that hormone was found long ago in the mouse, and recently in German Shepherd Dogs. Sporadic cases of stunted pups in that breed, and the occurrence of two in one litter, led to studies by Andresen et al. (1974) and Willeberg et al. (1975) which showed that the defect is a simple, recessive, autosomal character, as it is in the mouse.

Growth of the affected pups is retarded during the first two months and stops in a few weeks thereafter. The dwarf dogs are small, but well-proportioned, with partial retention of their puppy coats. In general, their appearance was described as "fox-like" (Figure 12-2). At sixteen months of age, one weighed only 5½ kg and her dwarf brother 14 kg. Another that ceased growing at 4 weeks and died at 12 weeks had a markedly underdeveloped adenohypophysis,and other abnormalities, including cleft palate.

From normal parents that all traced back to a common ancestor, the ratio of normal : dwarf in three litters was 14 : 4. Endocrinologists now believe that the growth hormone promotes growth by control of substances in the blood called somatomedins. Quantitative assays in four dwarfs revealed that all were abnormally low in somatomedins, unrelated controls were high, and some (but not all) of eight relatives of the dwarf dogs were intermediate. Presumably these last were heterozygotes, but phenotypically normal.

Subsequently, what was evidently the same recessive dwarfism was found in Carelian Bear-dogs, a breed that originated in the Carelian area of Finland (Andresen and Willeberg, 1976). From evidence that these had sometimes been crossed with German Shepherd Dogs at a period to which dwarfism in that breed has been traced, it seemed likely that one and the same mutation was responsible for the dwarfism in both breeds.

The pedigree of this condition shows 9 dwarfs and 10 proven carriers in five generations (Figure 12-3). It merits special study (and emulation) because of the excellent way in which it shows diagrammatically not only relationships among the carriers, but also complete records for the six litters of Generation V that each contained a dwarf. Omitting the two smallest of these and the four pups that died too early to be classified, the ratio was 23 normal : 7 dwarf, a remarkably close fit to the expectation of 22.5 : 7.5.

FIGURE 12-2
Three 17-month-old German Shepherd Dogs from the same litter. Two show pituitary dwarfism; the other is normal. [From Andresen et al., 1974.]

Autoimmune Thyroiditis

Autoimmunity is a condition in which the body forms antibodies against tissue components of one or more of its own organs. As a result, those organs are invaded by lymphocytes and (in varying degrees) are thus prevented from functioning normally.

Autoimmune thyroiditis was identified by Musser and Graham (1968) in about 12 percent of 981 Beagles in a large colony of that breed. Because most of these were descended from one affected female, and because there were many in some lines but none at all in others, it was concluded that the disease is hereditary. In 744 other Beagles that had been purchased, the incidence was below 4 percent, but those affected traced back to the same ancestors as did the affected animals in the colony.

In another Beagle colony, Fritz et al. (1970) found the incidence of autoimmune thyroiditis to be about 20 percent in dogs over one year of age. Studies of pedigrees and the high frequency in one inbred line showed that (as in the other colony) the thyroiditis was inherited.

In later investigations by Fritz et al. (1976), it was found that orchitis (inflammation of testes) was also present in many of the dogs that had thyroiditis. Among animals all descended from the same common ancestors, the incidence of lymphocytic thyroiditis was 85 percent, and of lymphocytic orchitis 65 percent. The latter condition

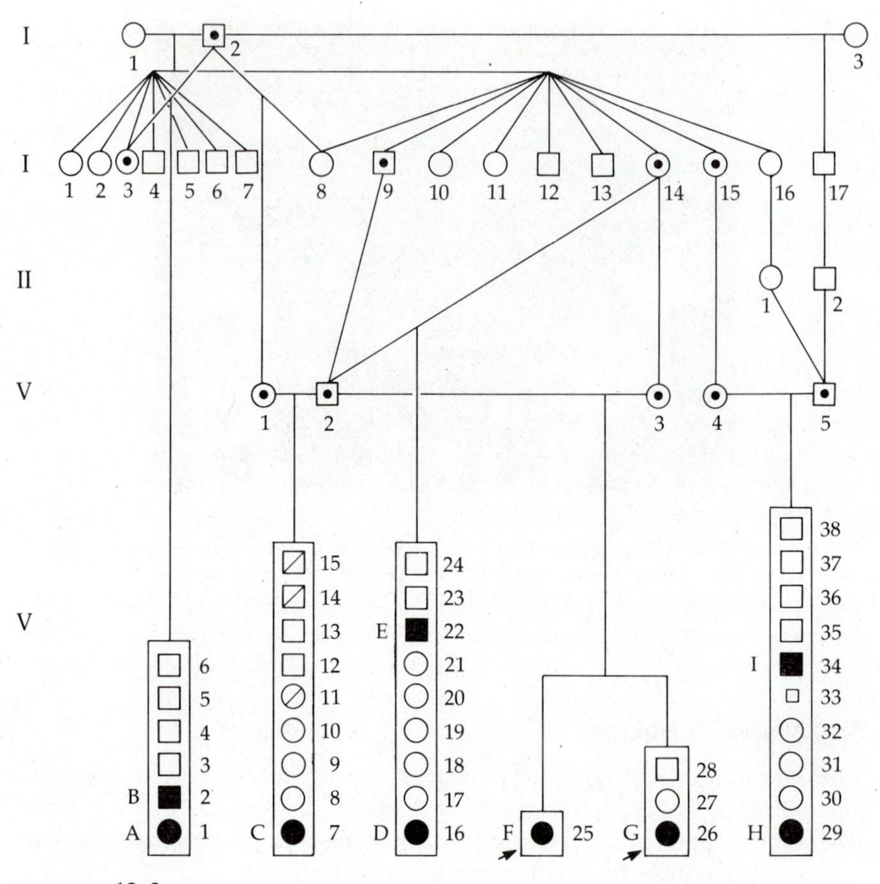

FIGURE 12-3
Pedigree for 9 Carelian Bear-dog dwarfs and 10 proven carriers. Numbers V-11, V-14, and V-15 died before diagnosis of dwarfism was possible, and Number V-33 was a stillborn male. Arrows point to the probands from which the other dwarfs and their ancestry were traced. [After Andresen and Willeberg, 1976.]

reduced significantly both size of testes and fertility of the affected dogs.

According to Musser and Graham, their dogs affected with autoimmune thyroiditis showed no physical evidence of that abnormality during the limited life span of the animals. Because most dogs in Beagle colonies are not kept there to ripe, old ages, it seems probable that clinical symptoms would be more evident in older animals.

The lesions found in autoimmune thyroiditis are similar to those in Hashimoto's disease, which is known to be hereditary in man. In the fowls affected with autoimmune thyroiditis studied by Cole et al. (1968), obesity was so marked that the strain affected became known as the Obese strain. In later years, Cole (personal communication, 1977) has (by continuous selection) raised the incidence of autoimmune thyroiditis in his White Leghorns to more than 96 percent.

The condition is clearly polygenic in the fowl, and may be the same in dogs. Because all of this section deals with Beagles, one must not conclude that autoimmune thyroiditis occurs only in that breed. It has been reported in others. Furthermore, the condition is more likely to be recognized in large colonies of animals than in small kennels, and there are probably 50 or more large Beagle colonies for every one of Great Danes or Chihuahuas.

Aberrant Metabolism

Utilization of the nutrients in our diet (and the dog's) depends on interactions by which various enzymes break down complex substances into simpler ones in forms that the body can use. Most of such metabolic processes proceed by a series of steps, each dependent on completion of all those before it, and each dependent on normal functioning of some particular enzyme. In a previous chapter, we considered three cases in which malfunction of essential enzymes resulted in pathological conditions in dogs: hemolytic anemia, deficiency of catalase, and gangliosidosis.

The last-named of these is one of a group sometimes termed *storage* diseases. In these, some intermediate product, still not broken down far enough to be utilized or eliminated, becomes accumulated in quantities great enough to cause disease. According to Koppang (1970), that can happen in three different ways: (1) The necessary enzyme may be lacking, or (2) its activity may be so diminished that the intermediate product which it should break down piles up, or (3) a normal enzyme mutates to an aberrant one that forms a product just abnormal enough to be useless. As it cannot be eliminated by normal mechanisms, that product accumulates and causes disease.

Neuronal Ceroid-lipofuscinosis

One such disorder in dogs is neuronal ceroid-lipofuscinosis. The word *ceroid* is a term for a fluorescent pigment. Some years ago, we would have known the similar disease in man as *juvenile amaurotic idiocy*, a hereditary, lethal disease attributed to a block in the breaking down of fats. According to Koppang (1970), at least 15 mutations are now known to cause such diseases (lipidoses) in man. These are differentiated by the kinds of fat products (lipids or lipoproteins) that accumulate in the body.

In Koppang's study of ceroid-lipofuscinosis in English Setters, he had over 30 cases from breeders and, during a period of eight years, 70 more from his own experimental matings (Table 12-1). The disease was clearly caused by an autosomal recessive gene when homozygous.

Clinical symptoms in the affected dogs became noticeable at 12 to 15 months, and included impaired vision along with mental degeneration shown by increasing dullness. The defect was detected earlier by

TABLE 12-1 Koppang's Experimental Matings Showing Inheritance of Ceroid-lipofuscinosis

Parents	Progeny	
	Normal	Affected
Affected ♂ × normal ♀ (not a carrier)	15	0
Carrier × normal (reciprocal)	18	0
Carrier × affected (reciprocal)	43	44
Carrier × carrier	104	25
Affected × affected	0	1

Source: Koppang (1970).

other dogs, which disliked those afflicted and in some cases even killed them. At 15 to 18 months, the amaurotic dogs became disoriented and unable to localize sounds. They showed difficulty in walking and gradually lost contact with their surroundings. By 18 months, some experienced muscular spasms, became cramped in the jaw, and clicked their teeth. Few of them survived to two years of age. Postmortem examination revealed extensive deposition of lipid granules in the nerve cells.

Subsequently, Patel et al. (1974) found that in the English Setters homozygous for neuronal ceroid-lipofuscinosis there was reduced activity of the enzyme PPD-peroxidase. Their scale for measuring activity of that enzyme put it at an average of 12.5 in seven normal controls, but at only 0.88 in thirteen homozygotes. From three heterozygotes, the average was six. This provides one more example in dogs of a hereditary disease in which the phenotypically normal animals that carry the causative gene can be detected by special tests in laboratories competent to make them. Such laboratories are not to be found in every village, but, fortunately, these storage diseases are not common. The breeder who does have to eliminate one of them (or any other that is detectable by test in heterozygotes) will find that it is better (and faster) to find a laboratory able and willing to make the appropriate tests than it is to identify heterozygotes by breeding tests.

A similar lipidosis was studied in German Short-haired Pointers by Karbe and Schiefer (1967) that they considered comparable to the late infantile type of amaurotic idiocy in man. They believed it to be sex-linked, but further evidence on that score is desirable. Koppang (1970) found that lesions of a lipidosis seen by him in Cocker Spaniels were not the same as those in the English Setters. Caution is evidently necessary in comparing lipidoses in dogs with those in man, and Koppang would go no further than to say that the disease he studied in English Setters was "very similar to, if not identical with the human juvenile amaurotic idiocy."

Part III

Breeds and Breeding

A Bit of Philosophy, and Some
Suggestions for Breeders

3 Breeds and Behavior

Breeds

"What is a breed?" This question is often asked by people who are not quite sure about differences between species, breeds, and varieties. All domestic dogs belong to the species *Canis familiaris.* There are at least nine species in the genus Canis, and taxonomists still argue a little (among themselves) about relationships of some of those nine and their rights to rank as distinct species. Three of the nine are jackals. The wild "yellow-dog dingo" of Australia and Kipling is *C. dingo* (Figure 13-1), the coyote is *C. latrans,* the wolf *C. lupus,* and the red wolf *C. niger.* There is also an Asiatic species, *C. sinensis.*

Although these nine species are spread around the world, at least six of them have one thing in common: they all have 78 chromosomes (Chiarelli and Capanna, 1973). It seems likely that when the other three are examined they too will conform to that number. Related species that have the same number of chromosomes can hybridize among themselves more successfully than species with differing numbers, so it is not surprising that hybrids from crosses of dog × coyote and dog × wolf are known to occur.

Descriptions and definitions of breeds and varieties within a species may differ. Long ago, when I had youthful aspirations of becoming a chicken expert, I was told that shape makes the breed and color makes the variety. In later years, the inadequacy of that definition became evident, and another was adopted. With the change of only one word, it could apply as well to dogs as to chickens: a breed is a group of dogs related by descent and breeding true for certain characteristics that the breeders agree to recognize as the ones distinguishing

FIGURE 13-1
The wild dingo *(Canis dingo),* which is widely distributed over the Australian continent. It has a broader head and stronger jaws than most domestic dogs. Because it attacks cattle and sheep, the dingo is considered by stockmen as a pest. In this picture, the mother anxiously guards her puppy. [Courtesy of Australian Information Service, New York.]

the breed. That definition should cover everything from Yorkshire Terriers to St. Bernards.

A variety is a subdivision of a breed. Differing colors give us several color varieties of Cocker Spaniels, while variations in size and coat provide varieties in other breeds. The American Kennel Club recognizes the Dachshund as a breed and three varieties within it (Smooth or Short-haired, Wire-haired, and Long-haired), also Miniatures in all three varieties. There are also dappled Dachshunds. Undoubtedly, devotees of any variety will consider it as a distinct breed in its own right and entitled to all the acclaim, privileges, and prizes accorded to any breed in the show-ring.

In some cases, breeders have to compromise a little on the stipulation that a breed should breed true to type. When it was shown that Blue Andalusian fowls are all heterozygous and produce (when two blues are mated together) blacks, blues, and blue-splashed whites in the ratio of 1 : 2 : 1, all that was necessary was a qualification in the standards to say that half the progeny of purebred Blue Andalusians should also be blue. Similarly, only half the foals from Palomino parents conform to the beautiful color of that breed, but mating together the two off-type kinds yields progeny that are all Palominos. In dogs, the merle gene makes heterozygous varieties in several breeds, but, because they are so attractive, we can afford to wink at that part of our definition of a breed (or variety) which says that it should breed true. In any case, blue merle Collies are clearly Collies, and not any other breed.

Collections of Genes

Let us think for a moment of dogs in general (and our pet Poodle in particular) not just as the oft-quoted man's best friend, but as collec-

(a)

(b)

FIGURE 13-2
Differing genes and different people make differing breeds. (a) Russian Wolfhound, or Borzoi; about 75 to 100 lbs. This is Ch. Flying Dutchman of Tamboer. [Courtesy of Lena Tamboer, Mahwah, New Jersey.] (b) Lhasa Apso, only about 10–11 inches high at the shoulder. [Courtesy of the American Kennel Club.]

tions of genes. Some of those genes that were accumulated during millions of years before the dog was domesticated determine that our dog should be born a dog—and not an elephant. Others determine that it will have characteristics common to the genus Canis, and still others make it *Canis familiaris* and not a dingo or a coyote. After the dog was domesticated, its owners began selecting for reproduction certain types that they preferred, and thus breeds were born.

As man spread over the world, he encountered variations in climate, in food supply, and in ways of life essential for survival. These, in turn, caused great diversity among ways in which dogs could be helpful to man. With continued selection for specific objectives, dog breeders in different walks of life developed breeds that excelled in some one of the many specialized canine skills. Hounds of different breeds were selected for skill in hunting deer, wolves, elk, foxes, coons, or other animals. How many modern lovers of Dachshunds know that they were originally bred to hunt badgers? Setters, Pointers, and Retrievers use their special skills to help hunters of birds. Terriers of various kinds were bred not only to be companions, but also to wage war on undesirable animals designated collectively as vermin.

Some breeds were developed to guard cattle or sheep. The Border Collie is justly famous as the shepherds' indispensable helper in herding sheep. Some breeds pull sledges in the Arctic; others, big and strong, were bred to pull carts in Europe. Most dogs can act as guards, and some breeds have been developed specifically for that purpose. One keeps unwanted strangers off barges in the Netherlands

(Keeshond), and another is said to be expert at sniffing out unwelcome intruders in Tibetan lamaseries (Lhasa Apso, Figure 13-2.) Many other examples of breeds developed for special skills will be known to most readers of this book.

To say that breeds of differing sizes and with differing skills were developed for special purposes is another way of saying that over the years the breeders preserved genes conducive to attainment of their ideals, and discarded those that were incompatible with those same ideals. Not all breeds were developed to be useful economically. Some that were bred originally to help the huntsman now provide entertainment as racers (Greyhound, Whippet).

Apart from the differentiation of breeds with specific skills, some of them have resulted chiefly from the desire of some breeders to have something different from the ordinary dogs of their neighbors. That urge can be satisfied by importing rare breeds from the far corners of the world[1] or by developing varieties of common breeds that differ in color or size. Why else should we have dappled Dachshunds, and miniature and toy varieties in so many breeds? Dog breeders are not alone in their understandable delight in the unusual. It has led chicken fanciers to produce hundreds of breeds and varieties that differ in color, shape, comb and size. Within the last 50 years, there has been a great proliferation of bantam breeds of the fowl, nearly all replicas on a smaller scale of standard breeds that are several pounds heavier.

Variation Under Domestication

Many years ago, Charles Darwin pointed out that there is far more variation in domestic species than in their wild ancestors. In the wild, every animal participates in the struggle for existence. By that struggle, a wild species becomes adapted to its environment, and a uniform type is established. Undoubtedly, mutations occur in wild animals as in domesticated ones. In nature, any mutation that reduces the biological efficiency of an animal is quickly eliminated. It may be a change that reduces ability of an individual to withstand a harsh environment, to resist disease, or to compete with other animals for food. The fittest survive—and reproduce their kind.

It is a different story under domestication. The environment becomes a sheltered one, or at least an improvement over conditions in the wild. Domestic animals no longer rely on genetic resistance to disease; in many cases, they are protected against exposure to disease or vaccinated to help them overcome it. The wild dingo has to compete with fellow dingoes and other species for food, but most domestic dogs are spared that problem. In fact (as the advertisements for dog food on television remind us), they can afford to be choosey about what they

[1] That Lhasa Apso again!

will or will not eat. Some mutations that might be a handicap in nature are not so under domestication and are often multiplied to form new breeds and varieties.

Over the centuries, and around the world, mutations in domesticated dogs have been numerous and varied. Also varied are the breeds that have arisen from those mutations. The observer marvelling at the diversity of breeds to be seen at any dog show might pause for a moment to reflect that each breed is a collection of genes, and that the differences among breeds show the differing preferences of breeders in various parts of the world for the mutations that have occurred since the first wild dogs were domesticated (Figure 13-2).

Many of those mutations affect polygenic characteristics such as size and conformation. Selection to accumulate genes for greater size resulted eventually in St. Bernards, Newfoundlands, and other large breeds. Selection to reject such genes and to accumulate those reducing size produced not only diminutive breeds like the Manchester and Yorkshire Terriers, but also resulted in standard, miniature, and toy sizes in other breeds. In Wales, breeders in Cardiganshire preferred rounded ears and long tails in their Corgis, but breeders in nearby Pembrokeshire selected for pointed ears and very short tails in theirs. Mutations that facilitated selection for these differing objectives were accumulated over the years (Figure 13-3).

The dog show that illustrates these variations in polygenic traits will also display the preferences of breeders for differing kinds of coats and other breed characteristics, some of which may have resulted from fewer mutations than those necessary to change size and shape. Coats can differ not only in color, but also to suit breeders who prefer hair that is long, short, wiry, or curly.

So go selection and the differentiation of breeds under domestication around the world. Diversity similar to that seen at a dog show can also be found at any large exhibition of chickens or pigeons. The wild mink in North America is black, but in domesticated minks over 50 mutations (or combinations of them) affecting color of the coat were recognized by the breeders as distinct, named varieties within the first half century of domestication.

Selection to Extremes

In most species of domestic animals, the characteristics that distinguish breeds are polygenic. Such characters can be "stretched" by selection beyond the limits of biological efficiency. For example, when breeders of turkeys found that broad-breasted birds brought higher prices than the older type with the prominent keel-bone of the wild turkey, they selected for broad, heavily muscled breasts. In a few years, they gave us turkeys so broad, so roly-poly, and so delightfully abnormal in conformation that they cannot reproduce as did their wild ancestors, and

(a)

(b)

FIGURE 13-3
Differing styles in Welsh Corgis: (a) Cardigan Welsh Corgi, with rounded ears and long tail; (b) Pembroke Welsh Corgi, with pointed ears and very short tail. [Both, courtesy of the American Kennel Club.]

are largely dependent on artificial insemination for their reproduction. Without it, they would probably become extinct. Similarly, judges of Hereford cattle (and breeders who sought to produce animals that pleased those judges) for some years gave preference to a short-legged, compact type of animal, but had to give up that preference when they found that it led to the production of unthrifty dwarf calves.

Dog breeders have gone to comparable extremes. As we have read earlier (Chapter 7), Dachshunds might have fewer slipped discs if their spines were shorter, and Boxers might have less spondylosis if theirs

FIGURE 13-4
English Bulldog, about 40–50 lbs. This is Ch. Westfield's Flying Colors. [Courtesy of Charles Westfield, Huntington, New York.]

were longer. Variations within one species in the number of vertebrae are inherited in fish, fowl, and swine, so there is reason to believe that selection concentrated on length of spine might reduce the amount of backache in the canine world.

Similarly, selection for extremely long and narrow (dolichocephalic) heads may account for the very high incidence of the assorted anomalies of the eye in Collies (Chapter 11). Selection in the opposite direction—for short noses and bashed-in faces (brachycephalic heads)—has gone to extremes in some breeds. For evidence on that score, one need only see the extremely brachycephalic English Bulldog that wins first prize in its class (Figure 13-4) and (at closer range) listen to the stertorous breathing that proclaims its pathological condition. That distress in breathing can cause not only excessive snoring but also loss of consciousness. They occur because the soft palate (in the roof of the mouth) is abnormally wide, long, and sometimes flabby. Distortion of the muzzle sometimes causes dental problems. Perhaps that extremely brachycephalic head was desirable when Bulldogs were bred to bite bulls on their noses, but in these modern times the English Bulldog is not expected to provide that sort of entertainment. The breed does attest to the skill of the breeders in stretching polygenic characters beyond the limits of biological efficiency. Now, when the English Bulldog can only reflect on the glorious traditions of its ancestors, perhaps some relief from its pathological state is in order.

Mutations at Random

While there is evidence that mutations can be induced experimentally by irradiation of different kinds and an assortment of chemical compounds, most mutations occur spontaneously. It is probable that all in the dog do so. Those that are good have been preserved to differentiate breeds and to enhance specific skills for man's use. The mutations causing bad effects are to some extent eliminated by natural selection or because the breeder rejects them as "off-type." Nevertheless, recessive mutations may crop out only rarely and can be passed along for generations before becoming recognized as hereditary defects. Many of these have been described in earlier chapters of this book.

Dog breeders see more of these undesirable mutations than do breeders of other domestic species because dog breeders do more inbreeding than do most animal breeders. Related animals have more genes in common than do unrelated animals. Inbreeding does not create bad genes, but it does bring together related heterozygotes that may carry the same bad gene, and thus produce homozygotes showing bird-tongue, ataxia, progressive retinal atrophy, or some other undesirable character. We shall have more to say about inbreeding in the next chapter. In this one, the chief point is that it brings to light the genetic junk that all dogs carry. Most people also carry genetic junk, but, because close inbreeding is taboo in *Homo sapiens,* we see somewhat less of our own bad genes than those of our dogs.

Mutations in Certain Breeds

Since most mutations occur entirely at random, they can occur in any breed. It is therefore no reflection on the good name or fame of any breed to find that some strains of it carry a hereditary defect. That point must be stressed. Breeders are so keen to preserve the good names of their pet breed that any bad news about it is sometimes "swept under the rug." In one case known to me, after a recessive defect had been found in a breed (which shall be nameless), a breeder unwilling to face the facts of life threatened to sue me, the veterinarian who diagnosed the defect, and Cornell University for defaming the breed!

In contrast with that dismal state of enlightenment, when the American Spaniel Club (1976) found that some genetic defects were prevalent in Cocker Spaniels, it published a Health Registry listing Cocker Spaniels that had been properly examined by competent agencies and found free of one or more of those faults. The defects were: cataracts, progressive retinal atrophy, deficiency of factor X, and hip dysplasia. Readers were warned that a dog not listed with respect to all four defects might be abnormal in those missing. It was interesting to find in a random sample of 150 dogs among the 715 that were listed, the proportions certified free of cataracts and PRA were 96 and 98 percent.[2] Only 40 percent of the dogs listed could boast an ample activity of factor X, but it is probable that, because deficiency of that essential element in blood clotting is apparently more common in some strains than in others, many animals were not tested. Similarly, since hip dysplasia is uncommon in Cocker Spaniels, it is probable that most breeders decided (understandably) to spend their money on items other than X rays. Only about 5 percent of the sample were certified to have sound hips.

Similarly, as we read in Chapter 10, when breeders of Otter Hounds found that genetic thrombasthenia was threatening the survival of their not-too-common breed, they appealed to competent authorities for help, and, in comparatively few years, practically eliminated the defect from show stock of the breed. Progressive retinal atrophy in Irish Setters was greatly reduced after breeders recognized that it is a genetic defect. Freudiger et al. (1973a) list the requirements specified by 14 breed clubs in Switzerland to reduce the incidence of hereditary hip dysplasia in the breeds that they represent. Some results of such precautions are shown in Chapter 14. Most dog breeders will know of other cases in which breed clubs have taken action to eliminate undesirable genetic defects.

Finally, let no breeder look askance at any breed other than his own in which genetic defects have appeared. His own breed probably carries genetic junk too, whether or not the undesirable genes have yet come to light. As the old saying has it, people who live in glass houses should not throw stones. An even older admonition applies (with change of one word) to breeds of dogs, as well as to the people to whom it was directed: "He that is without sin (fault) among you, let him first cast a stone at her" (St. John, 8:7).

Behavior

For two good reasons, comments in this book on genetic aspects of behavior will be brief. One reason is that, although much has been written about behavior in dogs, that part of it purporting to cover hereditary patterns has (in my opinion) not produced much convincing evidence of exact genetic bases for different kinds of behavior. The second reason is that most dog breeders who read this book are probably far more knowledgeable about canine behavior than I am.

---

[2]We are not told how many (or which) had tried their examinations and failed to pass.

Behavior is difficult to analyze in genetic terms, because so much of it depends on the environment (particularly the kind and amount of training) that it is hardly profitable to guess at how much is inheritance and how much is environment. As behavior is very difficult to measure in quantitative terms, most students of it have to resort to scales in which arbitrary judgements place the dog in one of half-a-dozen classes or more. How else could one study "inheritance of the tendency to be quiet while weighed" and emerge with the opinion that the tendency is recessive, perhaps dependent on two genes (Scott and Fuller, 1965, pp. 287–288)?

There is no doubt whatever that breeds differ in behavioral traits, or that there is some genetic basis for such differences as for other consistent differences among breeds. Pointer pups will point before getting any formal training in that art. Readers will know of other examples in which the dog seems to know by instinct the particular skill for which its breed was developed. Nevertheless, individuals differ within a breed, and some fail utterly to perform as the traditions of their breed demand. Even litter-mates can differ greatly. Not every Border Collie can respond to the intensive training that has made the breed famous for its ability to herd sheep. Disappointing dogs that are gun-shy do occur even in the best families of bird-dogs.

In many cases (if not all), the dogs that demonstrate superior skill in their particular craft are the ones that can best respond to training. A Labrador Retriever, Yogi, became an expert sniffer of contraband drugs and (in that capacity) a valued aid to the London police. He did not have drug-sniffing ancestors. He did have an uncanny sense of smell and an innate capacity to respond to the careful training by which his special art was perfected. Similarly, as all participants in obedience trials know, some dogs learn readily, others not.

It is futile (in my opinion) to try to guess how much of a dog's skill is genetic and how much is the result of training. Whether that skill is in driving sheep through a gate (Border Collie), retrieving ducks from murky waters (Chesapeake Bay Retriever), or rescuing people that would otherwise perish in Alpine storms (St. Bernard), both heredity and training play a part in its development.

Heritability

There are statistical techniques by which some geneticists believe that the separate roles of heredity and environment can be measured. Using those procedures (which I do not recommend to dog breeders), the operator thumps out on his calculating machine a coefficient of "heritability" that should lie between 0.0 and 1.0. If it is above 0.50, the trait is said to be highly heritable; that is, controlled by genes more than by the environment. Conversely, a heritability of 0.10 would indicate only slight genetic influence and almost overwhelming control by the environment. To determine coefficients of heritability, however, one would have to measure in mathematical terms the variations among individuals in the trait under study. That is easily done if one has only to count the eggs or weigh the milk, but what reader would venture to measure the degrees of perfection shown by the Border Collie, the Chesapeake Bay Retriever, or the St. Bernard in the behavior for which they are renowned?

Writings on Behavior

Without belaboring further the point that we know little about genetic aspects of behavior, but recognizing that some readers may have opened these pages in the hope of finding a lengthy discussion of that subject, it seems desirable to mention some other writings in which reading about behavior might prove more interesting than my comments.

1. The veteran dog breeder Leon Whitney (1972) listed 24 behavioral traits which he considered to be "natural" (i.e., independent of training) with which he himself had worked, and 12 more that had been studied by others. They ranged from trail-barking and "hound-drawl," through pugnacity and herding, to preferences among Dalmatians for different distances from the heels of horses. Detailed comments were given on many of the traits.
2. An excellent description of the breeding and training of guide dogs for the blind was given by Pfaffenberger (1963).
3. At the Jackson Laboratory in Bar Barbor, Maine, Scott and Fuller (1965) studied behavior in dogs for nearly 20 years. They devised tests to determine whether or not there were significant differences among breeds of dogs in such things as "escape activity," "bell response," body posture, tail wagging, and other indicators of behavior. Scales were devised for rating the dogs, and esoteric statistical procedures were used to interpret the results.

The breeds used were the Basenji, Beagle, Cocker Spaniel, Shetland Sheepdog, and Wire-haired Fox Terrier. Few knowledgeable dog breeders will be surprised to know that the investigators did find differences in behavior among the five breeds.

The only genetic studies reported were reciprocal crosses between Basenjis and Cocker Spaniels, with backcrosses to both parental breeds, and F_2 generations. From these, it was concluded that some behavioral traits (such as a tendency to fight the leash) were heritable but that others (obedience, tail wagging, etc.) seemed to be less so. Not all geneticists will be convinced. The book has an extensive bibliography on behavior in dogs and other animals.

4. Burns and Fraser (1966) reviewed much of what is known about behavior in dogs. The reader who is in a hurry might well read their two chapters on the subject before going on to the other books cited above. Genetic aspects of behavior are necessarily limited to differences among breeds, but these are well reviewed.

5. Finally, for information about genetic aspects of behavior in many different species (including man) readers should take a look at the recent book on that subject by McClearn and DeFries (1973). It has little to say about behavior in dogs, but gives a good exposition of the varied techniques, experiments, and statistical procedures used by investigators studying hereditary differences in behavior.

4 Selection and Breeding

In the lexicon of animal breeders, selection is the art of choosing the animals best qualified to beget the next generation. It is not in itself a science, but some knowledge of the science of genetics will help in perfection of the art.

This chapter is based in part on experience gained during my 40 years of poultry breeding, in part on principles of genetics, and, to a considerable extent, on an empathy with dog breeders that has been firmly established by many years of experience with their problems. Readers should remember that, whereas previous chapters of this book have dealt mostly with scientific principles and established facts, it is inevitable that this one must be based in large measure on the experience and beliefs of the author. It is hoped that readers will find it helpful.

There are only two kinds of selection. One is *individual selection,* also called *mass selection.* The other is *progeny-testing* and the selection based on the performance of *whole families* that it permits. Let us consider mass selection first.

Mass Selection

The method by which almost all dogs have been bred is mass selection. It is selection based on the phenotype and on the old saying that "Like begets like." The breeder chooses from among his available animals those which conform most closely to the ideal type that is desired. Perhaps he looks not only at the dog itself but also at the dog's pedigree to count (hopefully) the number of illustrious champions therein.

FIGURE 14-1
The massive St. Bernard and the tiny Chihuahua were differentiated by mass selection. The St. Bernard seems to share with a famous lady the thought that "We are not amused." [Courtesy of Gaines Dog Research Center, White Plains, New York.]

Like does beget like—within limitations. Dogs have puppies, not kittens; Beagles beget Beagles, not Bloodhounds. But champions do not always beget champions. Their chances for doing so are better when there are several generations of other champions behind them, but, as every dog breeder knows, pride of pedigree does not ensure pride in the progeny. Similarly, hens with several generations of good layers behind them can have poor layers among their daughters.

Admitting such limitations, let us recognize what mass selection has accomplished in the dog world. It is the method by which the many breeds have been established. Those breeds differ greatly in size, form, and function (Figure 14-1). They reflect the widely differing needs and whims of dog breeders around the world. They were not developed quickly. Most of the characteristics by which breeds differ are polygenic and dependent for full expression on the combined action of hundreds of genes, some of which have (by themselves) only very slight effect. One does not develop little Yorkshire Terriers or towering Great Danes in a few years, perhaps not even in a lifetime. One can only speculate about how long it took to produce the elongated faces of Collies and the shortened ones of bulldogs. Nevertheless, the distinguishing features of these four breeds—and of all others—were attained by mass selection.

Perfection in breed characteristics, and gradual changes in them to meet the whims of judges (and hence of breeders) are also dependent on many genes. Breeders will continue to use mass selection to attain that perfection. The difference between a champion and the dog next in line does not depend on 3 : 1 ratios.

One good reason why most breeders must use simple mass selection is that they have not got enough dogs to use the other method—progeny-testing. Before going on to explain the difference in the two procedures, let us consider a few examples of what mass selection is accomplishing in other species.

Mass Selection in Nature

Among all wild animals, there is a struggle for existence. The fittest survive. The fittest animals may be the fleetest, the best fighters, the most intelligent, those most able to reproduce, or those superior in some other way to the weaker individuals that succumb to the stresses of the environment or to competition within their own species. The fittest survive to reproduce their kind; the weaker animals are often eliminated before they can do so. The struggle by which some individuals survive while others cannot do so is commonly called *natural* selection. It is mass selection on a grand scale—and it is very efficient.

A good example of the efficacy of mass selection in wild animals was provided by the hardy oysters of Malpeque Bay. There, on the north shore of Prince Edward Island, in the Gulf of St. Lawrence, the oysters were nearly all killed in 1915 by an obscure disease that still bears no name other than Malpeque disease. A few survived, and ten years later persistent oystermen got four barrels of oysters. In 1926, the yield was up to 14 barrels, and in four more years only about one oyster in a thousand showed any signs of the disease. By 1938, the harvest was higher than in the years before the disaster of 1915.

That remarkable recovery of the species was accomplished by mass selection—survival and reproduction by individual oysters that were genetically resistant to the organism causing Malpeque disease. When experiments proved that the pathogenic organisms had not been eradicated, but that the oysters of Malpeque Bay had become genetically resistant to them, that resistant stock was used to accelerate development of genetic resistance in other areas to which the disease soon spread.

The fascinating story of how the oysters of Malpeque Bay conquered a serious disease in 12 years and then helped other oysters to do the same was told by Needler and Logie (1947) and was summarized later in condensed form (Hutt, 1958). The oysters of Malpeque Bay had in their armament a weapon of defence that few other species can match. It is the ability of single oysters to produce as many as 60 million potential little oysters in a year. That is why mass selection was so effective in the fast-working oysters of Malpeque Bay.

More familiar examples of rapid and effective mass selection in nature are provided by the house flies that quickly became resistant to the once-omnipotent DDT, and by the coccidia (of chickens) that become in a few years resistant to the best coccidiostatic drugs with which they are assailed. These, too, are small animals that multiply rapidly. They can therefore demonstrate in a few years the workings of natural selection. That same selection also works in elephants, but none of us is likely to see results of it in that species in our lifetime. We can see the operation of mass selection in the rabbits of Australia and elsewhere that have become in a few decades fairly resistant to the myxomatosis with which attempts were made to control them. Other examples of effective natural selection in animals from moths to cattle are given elsewhere (Hutt, 1964).

Natural selection also operates in domestic animals, but to a degree far less than that in nature. Most domesticated animals live in a sheltered environment and are protected against the struggle for existence by the care of their owners and the skill of the veterinary profession. Moreover, many domestic animals are not able to multiply their kind because they are excluded from reproduction by their owners.

Progeny-testing

Whereas mass selection evaluates prospective breeding stock by the phenotype, progeny-testing does so from estimates of the genotype. Those can be made only by inspecting the progeny. The easiest progeny-test is that for any defect known to be caused by a simple recessive gene. If a litter contains even one puppy afflicted with the bired-tongue defect, we know that both the sire and dam are heterozygous for the causative gene. If a litter from normal parents includes any male with hemophilia A, we know that the dam is heterozygous for the causative gene, and that the sire is free from it. He can be used safely in further breeding, but the dam cannot.

Appraisal of parental genotypes in dogs is not so easy when dealing with polygenic characters such as hip dysplasia and conformation. It is the ideal method for selection in cattle to raise milk production, or in domestic fowls to raise egg production, because performance of the daughters can be measured in quantitative terms as so many pounds of milk in one lactation period, or as the numbers of eggs laid in any arbitrary period of time.

In comparing progeny-testing with mass selection, one usually finds (in domestic animals) that the latter method will effectively raise production of milk, eggs, meat, or wool when first applied to unimproved stock. The rate at which it does so will depend in part on the degree of *selection pressure*. That pressure is obviously greater when the breeder reproduces from only the best 10 percent of the available animals than when he selects as a prospective sire any animal that is merely better than average (i.e., in the best 50 percent). The rate of improvement will also depend on the length of the interval between generations. If a poultry breeder waits until daughters under test have had a laying period of 300 days or more, he will have missed one breeding season and will be lucky if he can get three generations in five years. However, if he uses a short-time test and selects his best layers after a test of only 130 days, he can get a new generation every year.

Usually mass selection brings the stock to a plateau—a level beyond which little improvement can be made. At that point, progeny-test breeding properly applied will continue to raise productivity far beyond the level at which mass selection stopped giving results.

Few dog breeders can use progeny-tests effectively in their own kennels, for the simple reason that they do not have enough animals or facilities to do so. How many breeders of German Shepherds could test 10 young sires to find the ones which transmit the least hip dysplasia? That problem is aggravated by the fact that all the progeny would have to be kept a year or more before the hips could be accurately classified by X-ray examination.

Even though most dog breeders will have to continue using mass selection in their own kennels, by cooperative efforts of its members, a breed society could initiate progeny-tests and use them to reduce the incidence of whatever polygenic defects are recognized as being sufficiently prevalent in the breed to warrant concerted attempts at control. No one breeder is likely to test 100 Collie sires to find the 10 whose progeny have the fewest eye defects, but fifty breeders, each testing 2 sires, might do so. With proper precautions against inbreeding, those 10 good sires could then be utilized to disseminate their desirable genes through the kennels of the cooperating breeders.

Some general principles applicable in efficient progeny-testing are as follows:

1. That old adage, "the sire is more than half the flock" is correct. One male can sire 15 litters while one bitch is producing a single litter.
2. Select for testing the young sires most promising on the basis of pedigree, appearance, freedom from defects, good litter-mates, and any other possible criteria by which good prospects might be identified.
3. Mate the sire with unrelated bitches, preferably so chosen that all sires have comparable females.
4. Get from each sire under test enough litters to provide preferably 30 to 40 progeny, so that the proportions of desirable and undesirable pups among them can be expressed in percentages that are significant.
5. Include all progeny. Record the number stillborn or dying early. Keep all the live pups until the trait being studied can be measured.
6. Use the sires proven best as long as they can reproduce or until better ones are found.

Fallacies and Fetishes

The pros and cons of mass selection and progeny-testing in domestic animals have been discussed in greater detail elsewhere (Hutt, 1964, chap. 16). One point that should be made clear to dog breeders is that the pedigree behind an animal only offers hope that the most recent name in it will be a credit to the ancestors. There is no guarantee—no assurance that grandsons of champions will be champions. To be sure, a good pedigree is far better than none at all, but many animals that once enjoyed fame and fortune have subsequently (even years later) been found guilty of having spread through their breed some lethal gene, or hereditary defect. This has happened in dogs, horses, cattle—and in man.

To the geneticist, the pedigree provides means of measuring degrees of inbreeding, of tracing the descent of hereditary defects, and of identifying the animals that carry the genes causing those defects. For the animal breeder, the pedigree is similarly useful, and it also enables him to avoid close inbreeding. If it carries the names of illustrious champions, it serves the useful purpose of buoying up the breeder's hopes for the next generation. It is a far better guide than no pedigree at all. Apart from all these commendable features, the pedigree is usually essential for registering animals in the breed society's records.

With all these good things to recommend it, the pedigree has one drawback. It attracts those who worship pedigrees, be they breeders of horses, cattle, dogs, or any other domesticated species. For them, it becomes a fetish. My dictionary gives one definition of *fetish* as "irrationally reverenced." Let us not be pedigree-worshippers. A good progeny-test is worth more.

A common fallacy (particularly among poultry breeders) is that a *sib-test* is adequate for identification of superior breeding stock. The theory behind that belief is that, if full sisters prove to be good layers, their full brother (with exactly the same pedigree) is likely to beget daughters that will also lay well. With that belief, a progeny-test becomes unnecessary. For dog breeders, the corresponding illusion is that, since all pups in a litter have the same sire and dam, any one of them is equally as good as any other for breeding.

Unfortunately, the theory is not quite right. During 25 years, Dr. R. K. Cole and I tested many pairs and trios of full brothers in the effort to find superior sires. Even when mated to the same females, full brothers can yield entirely different progeny. In one trio thus tested, one male proved to be the best sire found in several years of testing (Hutt, 1964, p. 355). His full brothers were worse than mediocre. Sib-tests are good guides to the selection of males for testing. While a good male from a good family is a far better bet than a similar male from a poor family, only progeny-tests reveal which sires are the best.

Some breeders use progeny-tests only to find in good families young animals suitable for future breeding. That is a very limited use of progeny-testing. The real value of progeny-testing lies in finding superior sires and re-using them to disseminate their desirable genes through the flock, herd, or kennel.

Selection Against Simple Hereditary Defects

The simplest kind of breeding that the individual dog breeder may have to undertake is the elimination from his kennel of any defects caused by single genes.

It is easier to do so with dominant characters like lymphoedema than with recessive ones like bird-tongue (Chapter 1) because, in the former case, one just eliminates the animals that show the trait. Since most dominant characters are only incompletely so, it may be necessary (as with lymphoedema) to get expert help to detect the heterozygotes that show less than full expression of the dominant gene.

It takes more time to eliminate simple, recessive defects, especially when the homozygotes cannot survive to be used in test-matings for detection of carriers. In such cases, one must rely on proven heterozygotes for test-matings, and, as is made clear in Chapter 6, more offspring are then needed to prove that a suspect animal is not a carrier.

Fortunately the list is growing of hereditary defects in dogs that are detectable in heterozygotes by special laboratory tests. At present, most of such cases are defects in the blood, but, from the lengthening list of genetic defects in man that can be detected in phenotypically normal carriers, it seems probable that the corresponding list for the dog will also be extended. Obviously, an undesirable gene can be eliminated much more easily when laboratory tests make test-matings unnecessary. Breeders can rejoice (albeit a bit sardonically) over the fact that we can do more to reduce genetic junk in our dogs than in our own species.

It is difficult to cope with hereditary defects, whether dominant or recessive, that are not manifested until middle age or later. By that time, the causative gene will probably have been passed on to the next generation. An example is Huntington's chorea in man. In dogs, it is much simpler to deal with bird-tongue, which is evident at birth, than with epilepsy, which may not show up until many months later.

Selection Against Polygenic Defects

This is more difficult than with defects caused by single genes. It is also more difficult for small-scale breeders who raise only one or two litters a year than for those who maintain large kennels. The former are limited to mass selection; the latter can use progeny-tests, which are more efficient.

The small-scale breeder should not breed from dogs showing the defect and should also avoid using their litter-mates and parents. If possible, matings should be arranged with sires shown by progeny-test to transmit little or none of the defect. In large kennels, or whenever groups of breeders can arrange cooperative efforts, the program should be aimed at testing as many sires as possible to find the ones that transmit the defect in the least degree or not at all. These can then be used extensively.

The number of sires that should be tested cannot be stated arbitrarily. When only 5 are tested, the breeder might re-use the best 1 in 5. If 20 are tested, there should be at least 4 good ones, thus providing:

1. More sires to be available for cooperating breeders.
2. Hence less risk of inbreeding in subsequent generations.
3. A fair degree of selection pressure against the defect, since only the best 20 percent of those tested will be re-used.
4. Lastly (but by no means least) some assurance that if the best sire in the tests is subsequently lost by disease or accident, there will still be other good proven sires available.

As with single-gene defects, it is easier to do progeny-tests for characters dependent on polygenic inheritance when the trait can be identified at an early age than when complete diagnosis for a litter may not be possible until 8 to 12 months of age or later. Collie breeders should do well with the anomaly of the eye besetting that breed because it can be detected (with an ophthalmoscope) at 6 to 10 weeks of age. Progeny-tests for hip dysplasia are more difficult because, by the time the pups are old enough for definite diagnoses, a litter may be so reduced by disease, sales, or failure to get radiographs that accurate rating of the sire and dam is impossible. Complete litters are desirable in progeny-testing.

For defects in dogs that may show up only after several years (such as retinal atrophy, epilepsy, and cataracts in some breeds) progeny-tests are not very practical, but one can always hope that techniques will be developed by which such conditions can be diagnosed at younger ages.

Best results in progeny-testing are obtained when one can appraise the sire by his progeny in litters from several different females. If a single litter is exceptionally good, that result may be attributed to good genes from the sire, or from the dam, or from *specific combining ability* of the genes from both. However, if one sire begets six good litters from six different bitches, we can credit him with a good progeny-test. Equally reliable is a test of a female with several different males. Such matings can easily be made with sires in one year, but would probably require several years to be equally conclusive for a bitch. Five or six litters from one sire mated to different females should yield enough offspring for a dependable estimate of what the sire is transmitting. To say that only 12.5 percent of a sire's progeny show a polygenic defect means little when that percentage is based on only 8 pups; it becomes more significant when it is based on 30 to 40 offspring.

Eugenic Programs of Breed Societies

Numerous programs have been undertaken by various breed societies and kennel clubs to reduce the frequency of some specific genetic defect in some breed. It is beyond the scope of this book to review such programs. The usual requirements are that certificates showing freedom from the defect must be obtained before the dogs can be registered, shown, or entered for other competition. Sometimes similar certificates are required for importation. Although the requirements vary in different programs, they all have one objective in common—to discourage breeding from dogs that show the defect.

All of these programs have one more thing in common—they depend on mass selection for results. Elimination of the phenotypically unfit from reproduction is (hopefully) expected to reduce their number in succeeding generations. That it does, but, as we have read earlier, it is most effective in its first few years (or generations) of operation, less so thereafter.

Barnett (1970) credited a scheme for reducing progressive retinal atrophy with bringing the incidence of that defect in Border Collies (in Britain) from over 12 percent in 1965 down to less than 7 percent three years later. By 1973, the frequency of the defect (in Border Collies examined at the International Sheepdog Trials) had fallen below 2 percent (Barnett, 1976).

Similarly, Yakely (1972) conducted an educational program aimed at reducing the incidence of the Collie eye anomaly and reported that in a three-year period the incidence was reduced from 97 to 59 percent. The number of dogs involved was 370. In this case, mass selection against any form of the defect was supplemented to some extent by progeny-tests to identify animals that produced affected offspring.

Limitations of Such Programs

Lest breeders should expect too much too soon from such programs, it seems desirable to point out some of the limitations besetting them.

The most obvious of these is the fact that, when a defect is a simple recessive character, the number of animals showing it (i.e., homozygous) is far fewer in the breed (or kennel) than the number of heterozygotes. The former are usually easily recognized, but the carriers are phenotypically normal and hence not excluded (by most programs) from registration or from reproduction. Knowledgeable breeders will,

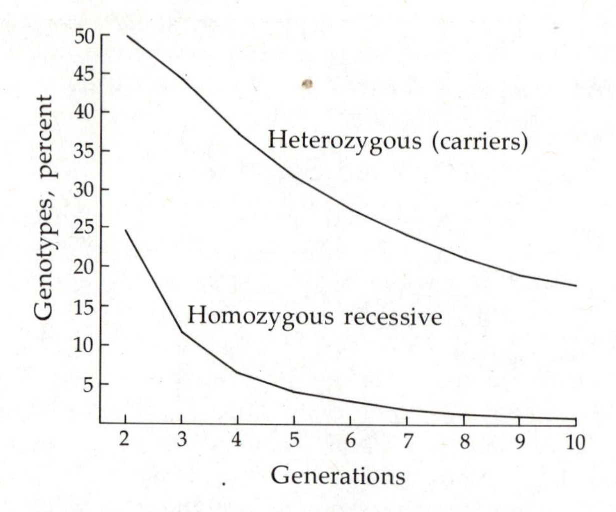

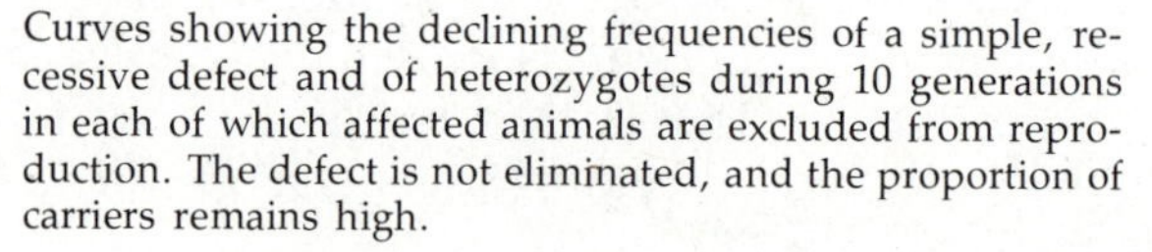

FIGURE 14-2
Curves showing the declining frequencies of a simple, recessive defect and of heterozygotes during 10 generations in each of which affected animals are excluded from reproduction. The defect is not eliminated, and the proportion of carriers remains high.

of course, carry on their own breeding tests to identify carriers and will stop breeding from them, but this is usually not required in the eugenic program of the breed society.

Too many breeders still think that if a hereditary defect is lethal, it will be self-eliminating, and that, similarly, all one need do about recessive viable defects is to exclude dogs showing them from reproduction. The fact is that even after 10 generations of such breeding there will still remain heterozygous carriers to pass the gene on to the next generation (Figure 14-2). If they happen to be champions, they can spread that gene far and wide.

A second handicap is the fact that some hereditary defects are not manifested until the dog is several years old. Progressive retinal atrophy is more easily reduced in Irish Setters (in which it is detectable at 6 to 10 months of age) than in Poodles, in which it may not appear until four to seven years (See Chapter 11). In the latter breed, many of the dogs will be reproducing long before the onset of their PRA. The same applies to other conditions with late onset, such as epilepsy and (in some breeds) cataracts. In some programs, registration is cancelled when a dog is later found to show the hereditary defect against which the program is aimed. This is scarcely more effective than locking the stable door after the horse is stolen.

A third shortcoming of some eugenic programs of breed societies is that, by concentrating their restrictions on dogs showing the undesirable abnormality, they fail to impress on the breeder the importance of excluding from reproduction the parents and litter-mates of the defective dogs. There are exceptions, and, in at least one society, registration is permitted only of dogs from approved parents.

Readers who feel that the limitations just considered are somewhat hypercritical and reflect, therefore, an unduly jaundiced attitude toward canine eugenics on my part, should also see the commendable exposition by Black (1972) of weaknesses in the scheme proposed in Britain for eradication of progressive retinal atrophy. From statistical analyses, and assuming an interval of three years between generations, he calculated that for the scheme to reduce the incidence of the defect to one in 10,000 would require only 1,080 years.

In spite of the limitations just listed, there are some hereditary defects with which programs of the breed society should be very effective. These are the conditions in which the causative gene is incompletely dominant, or codominant, so that it can be detected in both homozygotes and heterozygotes. Several such defects have been discussed earlier. For most of them, differentiation of the genotypes is made by laboratory tests of the blood.

Population Genetics

The special field of genetics called *population genetics* deals with the frequencies of genes in populations and with forces that influence changes in those frequencies. It is of particular interest to biologists who study evolution. Some of the principles and theories involved are also said to be applicable to the breeding of domestic animals when that is done on a large scale.

As most programs for breeding dogs are limited in size and scope, population genetics (in the opinion of the author) offers few applications to dog breeders except to those who maintain large colonies and supply dogs to laboratories for experimental studies. Readers who would like to know more about population genetics in animal breeding might consult the book on that subject by Pirchner (1969).

One principle of population genetics that could be useful to breed societies is the Hardy-Weinberg law. Its operation is explained in the paragraphs that follow.

Heterozygotes and the Hardy-Weinberg Law

When dealing with any simple recessive defect in a breed, there is usually no difficulty in identifying (by sight or by test) the animals that are homozygous for the causative gene. Even if they are eliminated from reproduction, the measures taken to reduce the incidence of the defect should consider the problem posed by the carriers of the gene, and their proportion among the phenotypically normal dogs of the breed. If they can be identified by some laboratory test, their frequency can be determined from tests of enough animals from different sources to comprise a representative sample of the breed. For example, Brown and Teng (1975) found thus that, among dogs tested by them, the proportion of Basenjis heterozygous for a deficiency of pyruvate kinase was 19 percent.

In most cases, the carriers cannot be so easily identified, but their proportions can be estimated by use of the Hardy-Weinberg law. It is so named after the two men who independently recognized it in 1908. The law is based on the fact that, when matings are entirely at random in a population, dominant and recessive genes of any one pair of alleles are maintained in the same ratio in successive generations.

With any pair of alleles, *A-a,* there can be three genotypes: *AA, Aa,* and *aa.* The problem is to estimate what proportion of the dogs having the normal phenotype are carriers of the recessive allele, *a.* To solve it with the aid of the Hardy-Weinberg law, we designate the frequencies of *A* and *a* algebraically as p and q. Since all the genes at each locus must amount to 100 percent (or 1.0), it follows that $p + q$ must equal 1.0. There can be $p \times A$ and $q \times a$, but the sum of both kinds must be 1.0. (Don't forget that every dog carries two genes at the locus concerned, hence can be *AA, Aa,* or *aa.*)

The law tells us that (with random mating) the frequencies of the three possible genotypes are

$$p^2\,AA + 2\,pq\,Aa + q^2\,aa$$

Anyone wondering how these frequencies were determined can make a little fourfold checkerboard (Punnett square), like the one we saw back in Chapter 2, but smaller. The male gametes, p *A* and q *a,* fertilize female gametes, also p *A* and q *a,* and the resulting combinations are as just given.

By substituting the proper values for p and q, we can determine the proportion of *Aa* genotypes in a random breeding population. We begin by finding q. If some condition known to be a simple recessive character shows up in the breed with a frequency of 1 per 100 dogs, then

$$q^2 = \frac{1}{100}$$

$$q = \frac{1}{10}$$

therefore

$$p = 1.0 - \frac{1}{10} = \frac{9}{10}$$

Applying these values for p and q in the Hardy-Weinberg formula,

$$p^2 + 2\,pq + q^2$$

we get

$$\left(\frac{9}{10}\right)^2 AA + \left(2 \times \frac{1}{10} \times \frac{9}{10}\right) Aa + \left(\frac{1}{10}\right)^2 aa$$

and the frequencies of the three genotypes are

$$81\,AA : 18\,Aa : 1\,aa$$

A point worth noting is that for every affected dog that we can see *(aa)* there are about 18 carriers *(Aa)* in the population (breed). These are (normally) indistinguishable among 99 dogs having the same, normal phenotype.

It is true that breeders do not make matings at random. They go on appearance, performance, and general desirability. However, so far as single, unseen, unknown genes that may be floating around in the breed are involved, the matings are essentially at random.

Obviously, the values for p and q used in the preceding example apply only to conditions that show up about once in every 100 dogs. For all others, they will vary according to the frequency of the condition in the breed or species. The Hardy-Weinberg law helps us to estimate the proportion of carriers, but, to do so, we must know the approximate frequency of the recessive condition in the population concerned. It cannot be applied to a small kennel, or to one from which some carriers have been eliminated by progeny-testing.

The law is useful for other purposes than estimating the frequencies of heterozygotes. If the actually observed proportions of the three genotypes do not fit the expectation of $p^2 + 2\,pq + q^2$, that proportion may have been upset by preferential mating, or by differences in viability of the three genotypes, but such matters are not yet shown to be of any importance to dog breeders.

Gene Frequency

The term gene frequency is commonly used to tell the frequencies of specific genes in a population (breed) when they can be determined. Thus, in the preceding example, there will be 200 genes at the *A-a* locus in 100 dogs. Frequencies are usually expressed as proportions of 1.0. In this case, they are

for *A*: $(81 \times 2) + 18 = 180$, or 90%, or 0.9
for *a*: $(1 \times 2) + 18 = 20$, or 10%, or 0.1

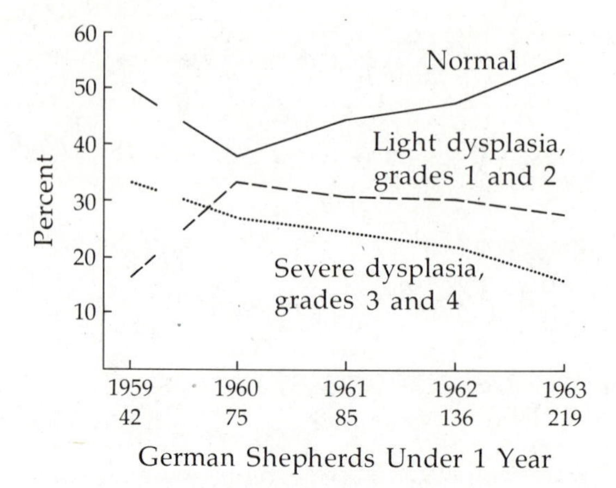

FIGURE 14-3
Incidence of hip dysplasia in random samples of German Shepherd Dogs in Sweden. The campaign against it began in 1959. Sizes of samples are shown below the years. [Data of S.-E. Olsson, 1963.]

Programs to Reduce Hip Dysplasia or Other Polygenic Defects

The Hardy-Weinberg law does not help one bit in programs aimed at reducing polygenic defects. The best known of these are the schemes employed in different countries, and in various breeds to reduce the incidence of hip dysplasia.

One of the earliest of these was that begun in 1959 by the Swedish Kennel Club. It required that a certificate of normal hips (diagnosed by X-rays) be provided for any German Shepherd Dog (1) to be awarded championship, (2) to compete in a get-of-sire class, (3) to be given a special prize for working dogs, or (4) if imported, to be registered in Sweden. Breeders were thus encouraged to breed only from dogs free of hip dysplasia.

Unfortunately, the incidence of hip dysplasia in Swedish German Shepherds did not decline as rapidly as had been hoped when the scheme was put into operation. The breeders became discouraged. To quote Schnelle (1973), "ten years of selective breeding not only failed to reduce the number of CHD [dysplastic] offspring, but also did not reduce the number of grade 2 or 3 (moderate or severe) cases." It is just possible that there was not as much "selective breeding" as was thought.

The requirements (as just given) did not put much selection pressure against hip dysplasia. They did not prevent anyone from using dysplastic dogs for breeding, and, in fact, concerned mostly those breeding German Shepherds for show purposes. Furthermore, they were based on mass selection—the common belief that if no dysplastic

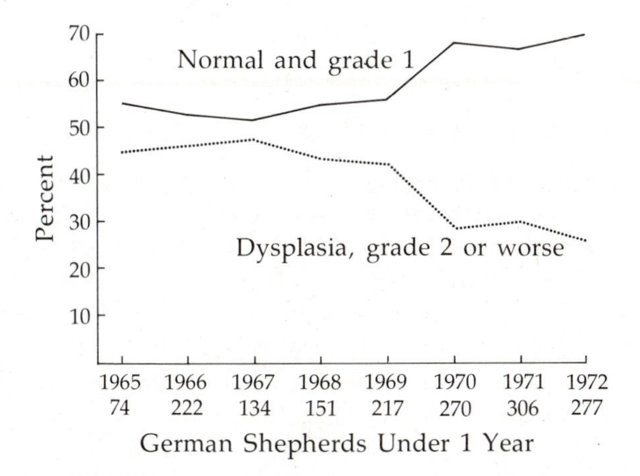

FIGURE 14-4
Hip dysplasia in German Shepherd Dogs in Switzerland during an eight-year campaign against it. Numbers of dogs examined (by X-rays) are shown below the years. [After Freudiger et al., 1973b.]

dogs became parents the defect would disappear. In view of the fact that matings of normal German Shepherd Dogs in Sweden at that time were yielding offspring in which 37.5 percent were dysplastic (see Table 5-1), that approach was hardly adequate.

Some years ago, from the data of Olsson (1963) on the incidence of hip dysplasia in random samples of Swedish German Shepherds during five years beginning in 1959, I plotted the frequencies of the different grades of dysplasia (Hutt, 1967). They showed a slow (but steady) increase in the proportion of dogs with normal hips, and a gradual decrease in the severe grades 3 and 4 (Figure 14-3). Undoubtedly the breeders had hoped for faster results, but, noting that there was little actual selection pressure against the defect, I was surprised to find that the frequencies of the two classes diverged as much as they did.

Somewhat better results from eight years' operation of a similar program in Switzerland were reported by Freudiger et al. (1973b). By 1972, among 277 dogs, the proportion with hips normal or grade 1 was up to 71.5 percent, while the proportion having grade 2 or worse was only 28.5 percent. Divergence of the frequencies in the two classes was attributed to the selection against hip dysplasia practised by the breeders (Figure 14-4).

Regulations in the German Democratic Republic against breeding from dogs affected moderately or severely with hip dysplasia are credited by Böhme et al. (1978) with reducing the proportion of German Shepherds having that condition from 44.3 to 12.4 percent from 1968 to 1975. Among dogs with hip dysplasia, the proportions with mild, moderate, or severe degrees of that abnormality were 6.7, 3.6, and 2.1 percent, respectively.

Progeny-testing and Polygenic Defects

It is not clear to what extent, if any, the breeders in these programs utilized the fact that with any polygenic defect there will be differences among sires in the proportions of their offspring that show that defect. In a program of breeding against hip dysplasia (or any other polygenic defect), the objectives should be (1) to identify the sires that transmit the least of it, and (2) to use such sires extensively, but to keep replacing them with still better ones as fast as the latter can be found. The ultimate aim is to get sires that have no dysplastic progeny.

This kind of breeding is not practicable for small kennels having at best only two or three sires. It can be utilized by a breed club, or by any group of cooperating breeders. To prevent inbreeding, superior sires found in any one kennel should not be bred back to their own progeny, but should be transferred in succession to other kennels so that their desirable genes become widespread in the breed.

It is highly desirable that the cooperating breeders or society should be able to test at least 20 to 30 sires annually, so that only the best 10 or 20 percent will subsequently be re-used as proven sires. When, from 30 young sires under test, only the best 5 are kept as proven sires, selection pressure is greater than when 10 are re-used.

Having used this system successfully for many years to raise genetic resistance to disease, egg production, and other desirable polygenic traits in fowls, I recommend it for coping with any polygenic traits in dogs. In our case, facilities permitted testing in each (annual) generation only about 10 to 15 young males per strain, and using also five proven sires per year in each strain (Cole and Hutt, 1973). Some of those good sires were used for four years. Had we been able to test more young males in each generation, and thus to exert more selection pressure in selecting proven sires for re-use, results would have come faster.

Breeders will find that when the program is in operation, the best prospects for testing among the available young males are not necessarily the sons of champions, but sons of the best proven sires.

This geneticist ventures to suggest that in breeding against hip dysplasia, breeders might be a bit less particular than usual about minor "faults" in the conformation of the hindquarters and mostly concerned about a low incidence of hip dysplasia. There have been suggestions that a major contributor to the problem is some kind of faulty conformation—slope of the croup, inadequate musculature, and so on. If so, changes in conformation may accompany reduced susceptibility to hip dysplasia. A good progeny-testing campaign that effectively lowers the incidence of dysplasia might possibly reveal what changes are desirable (if any) in conformation.

That Collie Eye Anomaly

If, as I believe, the Collie eye anomaly is a polygenic character, breeders seeking to eliminate it might well use exactly the same procedures as are recommended in the previous section for coping with hip dysplasia. In fact, the Collie breeders should find it easier to solve their problem than the breeders of German Shepherds (or other big dogs) have done with hip dysplasia. The difference is that one must wait a year or more for final diagnosis of hip dysplasia, but the eye defects can be detected (with an ophthalmoscope) within 6 to 10 weeks after birth. That means far better chances to classify every pup in a litter for eyes than for hips. Progeny-tests are of little use if based only on partial litters, but best when every pup can be classified.

In Chapter 11, we saw the records of two Collie sires, one of which had 60.7 percent of his progeny showing the eye defect, the other only 12.8 percent. By breeding from sires like the latter one (or better), it should not take too long to discover sires whose offspring are entirely free of the defect. As with hip dysplasia, it is suggested that the Collie breeders trying to find such sires might well forget about conformation of the head and concentrate on normal eyes. If the heads become a trifle wider, not everyone will be surprised.

Inbreeding

In many kennels, there is too much inbreeding. Often that happens because the breeder, having brought his dogs to the point at which they warrant some pride of achievement, is reluctant to introduce a new bloodline from another kennel, lest it should bring some kind of imperfection. Sometimes the inbreeding is done because the breeder hopes that his champions will beget more of the same.

Inbreeding does not of itself create any undesirable genes, but it will bring to light any of them that have been lurking, undetected, in good dogs of fine appearance. That happens because related animals are likely to have many genes in common, and, among these, there may be some recessive genes that induce defects in dogs homozygous for them. Such genes are safely hidden in heterozygotes, but, when such carriers are mated together, about a quarter of the progeny will be homozygous and can show the defect. Even with polygenic defects, matings of related animals can result in accumulations of the causative genes—and resultant greater expression of the defect.

There are differing degrees of inbreeding. The most intense is self-fertilization, which is common enough in plants but in animals is found only in some of the lower forms. Many breeders know that inbreeding is risky, but think that "line breeding" is safer. By inbreeding, they mean brother × sister, or parent × offspring; by line breeding, the mating of relatives more distant. Actually, mating within one breed is to some extent inbreeding, because dogs of one breed have in common the genes that induce the essential characteristics of the breed.

Both inbreeding and line breeding have the same effect (that is, they increase homozygosity), but line breeding does it much more slowly, and permits the breeder to stop the procedure at the first sign of trouble. With closer inbreeding, associated problems (infertility, small litters, outcropping of bad genes) may show up suddenly and then be difficult to eliminate.

Wright's Coefficient of Inbreeding

Because the degrees of inbreeding vary, it is desirable to have some standard measure by which it can be determined. Wright's coefficient, the one most commonly used, is

$$F_X = \Sigma \left(\frac{1}{2}\right)^{n + n' + 1} (1 + F_A)$$

where F_X = coefficient of inbreeding of the animal X
n = number of generations from the sire of X back to some ancestor common to both sire and dam
n' = number of generations from the dam of X back to that same common ancestor
Σ = summation of the separate contributions of each different common ancestor
F_A = coefficient of inbreeding of the common ancestor (A) when that animal is itself inbred

What Wright's coefficient measures is the degree of relationship between the sire and the dam of X. That will depend on how many ancestors are common to both sire and dam, and on how far back those common ancestors are. The separate contributions of each are added up to determine F_X. The figure 1/2 comes from the fact that the genes coming down to X from a common ancestor are halved with each generation that separates them.

In many cases, the common ancestor is not inbred, so the formula can then be simplified to

$$\Sigma \left(\frac{1}{2}\right)^{n + n' + 1}$$

The following two examples shown how the formula is used:

1. Mating of brother × sister to produce X

X — A {C, D}; B {C, D}

The common ancestors are C and D. The number of generations from A and B back to each is one.

Contribution from C: $\frac{1}{2}^{1+1+1} = \left(\frac{1}{2}\right)^3 = \frac{1}{8} = 0.125$

Contribution from D: Same as for C $= \underline{0.125}$

$F_X = \Sigma = 0.250$

X is 25 percent inbred.

2. Mating of half-sibs to produce X

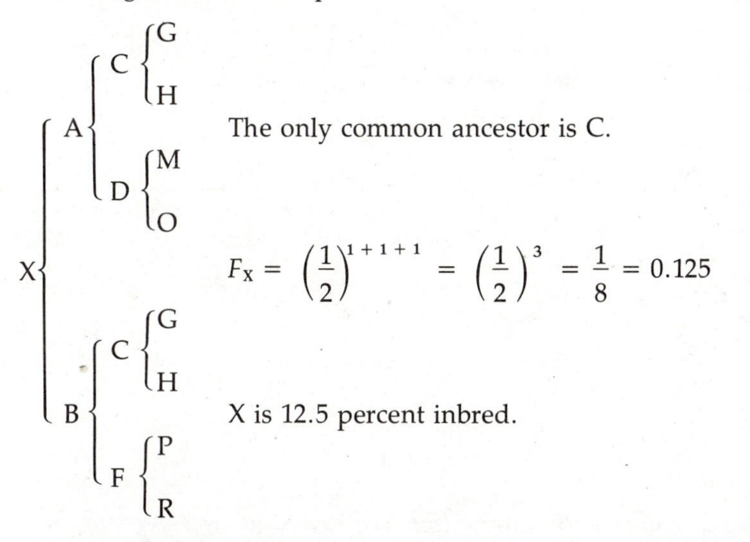

Any breeder attempting to work out a coefficient of inbreeding should remember that one counts the number of generations back to the common ancestor—not from X, but from the sire and dam of X. The farther back we go for that common ancestor, the smaller is its contribution to the inbreeding of X (Table 14-1).

The reason for including the calculation of Wright's coefficient of inbreeding in this book is *not* to exhort readers to try it on the pedigrees of some of their favorite dogs, but, rather, just to let readers know what it is all about when they read (somewhere else) that a certain dog (or a breed) is 20, 30, or 40 percent inbred. (The average for a

TABLE 14-1 Contributions of Some Common Relationships to the Coefficient of Inbreeding

Number of Generations Back to the Common Ancestor		
Behind One Parent	Behind the Other Parent	Contribution to the Coefficient of Inbreeding
0	1	$(1/2)^2$ = 0.25
1	1	$(1/2)^3$ = 0.125
1	2	$(1/2)^4$ = 0.0625
2	2	$(1/2)^5$ = 0.0312
2	3	$(1/2)^6$ = 0.0156
3	3	$(1/2)^7$ = 0.0078
3	4	$(1/2)^8$ = 0.0039
4	4	$(1/2)^9$ = 0.0019

breed can be determined from the coefficients of inbreeding for a representative sample of that breed.)

One reason why breeders don't need to measure inbreeding exactly is that there is no set figure above which it is a risky business and below which it is safe. In general, the higher the inbreeding, the greater is the risk. Apart from the genetic defects that crop out with inbreeding, there is usually (at higher levels of it) some decline in viability and ability to reproduce. In a large colony of Beagles, Rehfeld (1970) found the proportions of pups born that survived at least 10 days to be

Number of Pups	Coefficient of Inbreeding	Survived (Percent)
636	0.008–0.186	75
565	0.251–0.785	51
39	over 0.673	25

A fairly safe rule of thumb for dog breeders is never to mate brother × sister, or parent × offspring. Matings of uncle × niece or aunt × nephew are risky; so is the combination: male × half-sisters. A mating of first cousins is less so. In large kennels, contrary to common belief, it is not essential to introduce new stock periodically to hold down the level of inbreeding so long as (1) several sires are used in each generation; (2) the breeding stock is chosen without deliberate inbreeding, with all sires contributing about equally to the next generation; and (3) the number of females greatly exceeds the number of males. Wright (1931) calculated that, under such conditions, the heterozygosity in a population is reduced in each generation by $1/8n$, where n is the number of males used. If 3 males be used, the heterozygosity is reduced by only about 4.2 percent in each generation; when 10 are used, the figure is only 1.2 percent. Such a slow approach to homozygosity is unlikely to cause trouble in any kennel.

Resistance to Disease

We have already considered many cases of genetic resistance to disease in dogs. The gene *(H)* that protects heterozygous females *(Hh)* from having hemophilia is a gene for resistance to disease. So are the normal alleles of many recessive deleterious genes that induce other genetic defects. One may well ask, "Are there also genes that protect against infectious diseases caused by bacteria, viruses, protozoa, or other pathogens?" Yes, there are, but we know very little about genetic resistance to infectious diseases in dogs.

Twenty years ago, I reviewed what was then known about genetic resistance to disease in domestic animals (Hutt, 1958). Suffice it to say that whenever genetic differences among individuals in resistance to disease were adequately sought, such differences were found. In some cases, they were utilized to develop resistant strains. In domestic animals, however, genetic resistance is little utilized, and the prevailing measures for control consist in large part of sanitation, vaccination, medication, and periodically (in some species) slaughter to prevent a newly introduced disease from spreading. It is different in wild animals, as we have seen in Chapter 13.

Most dog breeders already have enough objectives without attempting to breed dogs resistant to various infections. Among those objectives, high priority must be given to eliminating genetic defects such as those described in this book. It is also plain common sense not to breed from dogs that are unusually susceptible to infections. Beyond that, few breeders have enough animals to justify attempts to attain genetic resistance to infectious diseases. Exceptions are the breeders who maintain large colonies of dogs and supply them to drug companies and hospitals for research. Ways in which they can raise genetic resistance to disease have been given earlier in this chapter.

Coliform Enteritis

Genetic differences in resistance to infectious disease show up when there is an outbreak that spreads through any large flock, herd, or kennel in which sire-families or dam-families can be identified. They are also evident when adequate samples of breeds that differ in resistance are all exposed to some disease at one time and in one environment. For example, Fox et al. (1965) found that, in a colony of over 250 dogs, coliform enteritis (dysentery caused by bacteria of the *Escherichia coli* group) was restricted to the 62 Basenjis, while, in four other breeds on the same premises at the same time, only sporadic cases occurred.

That difference did not mean that only the Basenjis were exposed. Several different types of the coliform bacteria were isolated from the Basenjis, in which they caused diarrhea and mortality of puppies, but the same types were also isolated from Cocker Spaniels, Beagles, Shelties, and Fox Terriers, which were unaffected by them. Two of the more resistant Basenji bitches produced similarly resistant progeny.

Fox et al. (1965) attributed the susceptibility of the Basenjis to the fact that they were highly inbred, and the other breeds not. It seems more likely to me that the Basenjis, which have been bred in the United States only since 1941 (American Kennel Club, 1968), had encountered one or more strains of *E. coli* to which their ancestors in Africa had never been exposed, and to which no genetic resistance had been built up by natural selection. Similarly, measles and chicken pox, which in most parts of the world are not considered serious diseases, wrought havoc among the Indians of North American when first introduced by invaders from Europe. The Europeans had undergone thousands of years of natural selection—the Indians, none.

Distemper

Quantitative data on the incidence and severity of distemper in different breeds of dogs are not available, but experienced breeders and veterinarians are convinced that genetic differences in resistance to that disease do exist. During his long experience in raising many breeds of dogs, Whitney (1972) had to cope with over 500 cases of distemper. He found Bloodhounds to be particularly susceptible, and those of English origin more so than the American ones. Oddly enough, his "bullhounds" from the cross Bloodhound × Bull Terrier were markedly resistant. A French veterinarian, Cazenave (1950), noted that in his practice Pomeranians and terriers were affected only slightly, other breeds more severely, but the Malinois, Groenendael, and German Shepherd worst of all. He believed that differences among breeds in susceptibility to distemper were in some measure responsible for variations in the efficacy of vaccination. Unfortunately, he gave no data.

Other evidence that breeds differ in response to vaccination for distemper was found by Baker et al. (1962), who assayed in several breeds the titre (amount) of antibodies against distemper at appropriate periods after vaccination against that disease. They found significantly higher titres (i.e., more resistance) in Labrador Retrievers and Cocker Spaniels than in Beagles and Pointers.

While the differences among breeds in resistance to distemper are genetic differences, it should be made clear that within any breed there can be differences among individuals. Although some may be resistant, others will be susceptible, and, as most owners of dogs know, the safest policy is not to rely on genetic resistance, but to vaccinate.

A Kind Word for Mongrels

It is to be hoped that no breeder of purebred dogs will mind if this book carries a message of good cheer to the mongrels. After all, they do comprise a big proportion of the dogs in the world.

Mongrels have varying degrees of a commodity now much sought by breeders of cattle, sheep, swine, chickens, and other domestic animals. It is hybrid vigor, now commonly referred to by geneticists as *heterosis.* It is the somewhat intangible extra vim, vigor, and vitality that lets good hybrid White Leghorns lay about 25 eggs more per year than their non-hybrid parental strains. It enables crossbred beef cattle to reach market weight sooner (and on less feed) than could their purebred parents. It helps broilers get to market weight at about 8 weeks of age instead of the 12 weeks once required—and they do so on less feed. The classic example of hybrid vigor in domestic animals is provided by that hardy hybrid—the mule.

So far not much deliberate crossing of breeds of dogs is done except for specific purposes, such as to get fairly quiet ones for certain types of experimental surgery, or (as animal breeders do the world over) to produce a new breed.

Nevertheless there are a few straws in the wind which suggest that after Armageddon the mongrels are more likely to inherit what is left of the earth than are the pure breeds. Some of these have been cited earlier. Patterson's (1968) mixed breeds had less than a third as many heart defects as had his pure breeds. Whitney's (1972) crossbreds (from Bloodhound × Bull Terrier) were highly resistant to distemper, and Hayes (1974) found that crossbreds had a very low risk of inguinal hernia.

Perhaps the mongrels have other additional expressions of hybrid vigor that will be discovered when their kind gets more study. The common belief that they can cope better than purebred dogs with the fleas and lice that beset the species has yet to be proven. At any rate, a crossbred dog may be called a mongrel, or a cur, but we might remember to consider him also as an interesting piece of hybrid vigor.

References

Aguirre, G. 1976. Inherited retinal degenerations in the dog. *Trans. Amer. Acad. Ophth. Otol., 81:* 667–676.

Aguirre, G. D., and L. F. Rubin. 1972. Progressive retinal atrophy in the Miniature Poodle: An electrophysiologic study. *J. Amer. Vet. Med. Assoc., 160:* 191–201.

Ahmed, I. A. 1941. Cytological analysis of chromosome behaviour in three breeds of dogs. *Proc. Roy. Soc. Edinburgh,* Sec. B, *61* (Part I): 107–118.

Allison, A. C., W. ap Rees, and G. P. Burn. 1957. Genetically-controlled differences in catalase activity of dog erythrocytes. *Nature, 180:* 649–650.

American Kennel Club. 1968. *The Complete Dog Book: The Official Publication of the American Kennel Club.* New York: Doubleday.

American Spaniel Club. 1976. *First Annual Health Registry.* (Valid through 3/31/77.) Vol. 1. Available from Mrs. E. H. Durland, Ellsworth Road, R.D. 1, Baldwinsville, N.Y. 13027.

Andersen, A. C., and F. T. Shultz. 1958. Inherited (congenital) cataract in the dog. *Amer. J. Pathol., 34:* 965–975.

Andersen, H., B. Henricson, P.-G. Lundquist, E. Wedenberg, and J. Wersäll. 1968. Genetic hearing impairment in the Dalmatian dog. *Acta Oto-Laryngologica,* Suppl. 232, pp. 1–34.

Andresen, E. 1977a. Haemolytic anaemia in Basenji dogs. *Hereditas, 85:* 211–214.

Andresen, E. 1977b. Haemolytic anaemia in Basenji dogs. 2: Partial deficiency of erythrocyte pyruvate kinase (PK; EC 2. 7. 1. 40) in heterozygous carriers. *Animal Blood Groups and Biochem. Genet., 8:* 149–156.

Andresen, E., and P. Willeberg. 1976. Pituitary dwarfism in Carelian Bear-Dogs: Evidence of simple, autosomal recessive inheritance. *Hereditas, 84:* 232–234.

Andresen, E., P. Willeberg, and P. G. Rasmussen. 1974. Pituitary dwarfism in German Shepherd Dogs: Genetic investigations. *Nord. Vet.-Med., 26:* 692–701.

Asdell, S. A. 1966. *Dog Breeding.* Boston: Little, Brown. IX + 194 pp.

Ashton, N., K. C. Barnett, and D. D. Sachs. 1968. Retinal dysplasia in the Sealyham Terrier. *J. Path. Bact., 96:* 269–272.

Averill, D. R., Jr., and R. T. Bronson. 1977. Inherited necrotizing myelopathy of Afghan Hounds. *J. Neuropathol. and Exper. Neurol., 36:* 734–737.

Baker, J. A., D. S. Robson, B. Hildreth, and B. Pakkala. 1962. Breed response to distemper vaccination. *Proc. Anim. Care Panel, 12:* 157–162.

Barnett, K. C. 1965. Canine retinopathies. 2: The Miniature and Toy Poodle. *J. Small Anim. Pract., 6:* 93–109.

Barnett, K. C. 1969a. The Collie eye anomaly. *Vet. Rec., 84:* 431–434.

Barnett, K. C. 1969b. Genetic anomalies of the posterior segment of the canine eye. *J. Small Anim. Pract., 10:* 451–455.

Barnett, K. C. 1970. The British Veterinary Association/Kennel Club Progressive Retinal Atrophy scheme. *Vet. Rec., 86:* 588–592.

Barnett, K. C. 1976. Comparative aspects of canine hereditary eye disease. *Advances Vet. Sci. Compar. Med., 20:* 39–67.

Barnett, K. C., G. R. Björck, and E. Koch. 1970. Hereditary retinal dysplasia in the Labrador Retriever in England and Sweden. *J. Small Anim. Pract., 10:* 755–759.

Barnett, K. C., and W. L. Dunn. 1969. The International Sheep Dog Society and progressive retinal atrophy. *J. Small Anim. Pract., 10:* 301–307.

Barnett, K. C., and G. C. Knight. 1969. Persistent pupillary membrane and associated defects in the Basenji. *Vet. Rec., 85:* 242–249.

Bernstein, F. 1929. Variations-und Erblichkeitsstatistik. In *Handbuch der Vererbungswissenschaft.* Bd. I. [Vol. 1] 1C. Berlin: Gebrüder Bornträger. 96 pp.

Bielfelt, S. W., H. C. Redman, and R. O. McClellan. 1971. Sire- and sex-related differences in rates of epileptiform seizures in a purebred Beagle dog colony. *Amer. J. Vet. Res., 32:* 2039–2048.

Bjerkås, I. 1977. Hereditary "cavitating" leucodystrophy in Dalmatian dogs. *Acta Neuropathologica, 40:* 163–169.

Bjork, G., S. Dyrendahl, and S.-E. Olsson. 1957. Hereditary ataxia in Smooth-haired Fox Terriers. *Vet Rec., 69:* 871–876.

Björck, G., W. Mair, S.-E. Olsson, and P. Sourander. 1962. Hereditary ataxia in Fox Terriers. *Acta Neuropathologica,* Suppl. 1, pp. 45–48.

Black, L. 1972. Progressive retinal atrophy. A review of the genetics and an appraisal of the eradication scheme. *J. Small Anim. Pract., 13:* 295–314.

Böhme, R., E. Schönfelder, and S. Schlaaf. 1978. Prognostische Untersuchungen zur Verbreitung der Hüftgelenksdysplasie beim Deutschen Schäfer-hund in der D.D.R. *Monatshefte f. Veterinärmedizin,* 33: 93–96. (Cited from *Anim. Breed. Abst., 46:* p. 595, 1978.)

Bovee, K. C., and S. Segal. 1971. Canine cystinuria and cystine calculi. Paper presented at 21st Gaines Vet. Symposium, October 20, 1971, Ames, Iowa, pp. 3–7.

Briggs, L. C., and N. Kaliss. 1942. Coat color inheritance in Bull Terriers. *J. Hered., 33:* 222–228.

Brinkhous, K. M., and J. B. Graham. 1950. Hemophilia in the female dog. *Science, 111:* 723–724.

Brinkhous, K. M., P. D. Davis, J. B. Graham, and W. J. Dodds. 1973. Expression and linkage of genes for X-linked hemophilias A and B in the dog. *Blood, 41:* 577–585.

Brown, R. V., and Y. Teng. 1975. Studies of inherited pyruvate kinase deficiency in the Basenji. *J. Amer. Anim. Hosp. Assoc., 11:* 362–365.

Bull, R. W. 1974. New knowledge about blood groups in dogs. *Gaines Research Symposium,* pp. 29–30.

British Veterinary Association. 1955. Cryptorchidism in the dog. *Vet. Rec., 67:* 472–474.
Burns, M., and M. Fraser. 1966. *Genetics of the Dog.* (2nd ed.). Edinburgh: Oliver & Boyd; Philadelphia: J. B. Lippincott. X + 230 pp.
Burstone, M. S., E. Bond, and R. Litt. 1952. Familial gingival hypertrophy in the dog (Boxer breed). *Arch. Pathol., 54:* 208–212.
Carrig, C. B., and A. A. Seawright. 1969. A familial canine polyostotic fibrous dysplasia with subperiosteal cortical defects. *J. Small Anim. Pract., 10:* 397–405.
Cawley, A. J., and J. Archibald. 1959. Ununited anconal processes of the dog. *J. Amer. Vet. Med. Assoc., 134:* 454–458.
Cazenave, M. 1950. Influence de la race sur la maladie de Carré. *Rev. Méd. Vétérinaire, 101:* 63–65.
Cheville, N. F. 1968. The gray Collie syndrome. *J. Amer. Vet. Med. Assoc., 152* (Part I): 620–630.
Chiarelli, A. B., and E. Capanna (eds.). 1973. Cytotaxonomy and Vertebrate Evolution. London and New York: Academic Press. XVI + 783 pp.
Clifford, D. H., E. D. Waddell, D. R. Patterson, C. F. Wilson, and H. L. Thompson. 1972. Management of esophageal achalasia in Miniature Schnauzers. *J. Amer. Vet. Med. Assoc., 161:* 1012–1021.
Clough, E., R. L. Pyle, W. C. D. Hare, D. F. Kelly, and D. F. Patterson. 1970. An XXY sex-chromosome constitution in a dog with testicular hypoplasia and congenital heart disease. *Cytogenetics, 9:* 71–77.
Cockrell, B. Y., R. R. Herigstad, G. L. Flo, and A. B. Legendre. 1973. Myelomalacia in Afghan Hounds. *J. Amer. Vet. Med. Assoc., 162:* 363–365.
Cole, R. K., and F. B. Hutt. 1973. Selection and heterosis in Cornell White Leghorns: A review with special consideration of interstrain hybrids. *Anim. Breed. Abstr., 41:* 103–118.
Corley, E. A., T. M. Sutherland, and W. D. Carlson. 1968. Genetic aspects of canine elbow dysplasia. *J. Amer. Vet. Med. Assoc., 153:* 543–547.
Crawford, R. D. 1970. Epileptiform seizures in domestic fowl. *J. Hered., 61:* 185–188.
Crawford, R. D., and G. Loomis. 1978. Inheritance of short coat and long coat in St. Bernard dogs. *J. Hered., 69:* 266–267.
Croft, P. G. 1968. The use of the electro-encephalograph in the detection of epilepsy as a hereditary condition in the dog. *Vet. Rec., 82:* 712–713.
De Boom, H. P. A. 1965. Anomalous animals. *South African Journal of Science, 61:* 159–171.
De Lahunta, A., and D. R. Averill. 1976. Hereditary cerebellar cortical and extrapyramidal nuclear abiotrophy in Kerry Blue Terriers. *J. Amer. Vet. Med. Assoc., 168:* 1119–1124.
De Wit, C. D., N. A. C. J. Coenegracht, P. H. A. Poll, and J. D. v. d. Linde. 1967. The practical importance of blood groups in dogs. *J. Small Anim. Pract., 8:* 285–289.
Dodds, W. J. 1970. Canine von Willebrand's disease. *J. Lab. Clin. Med., 76:* 713–721.
Dodds, W. J. 1973. Canine factor X (Stuart-Prower factor) deficiency. *J. Lab. Clin. Med., 82:* 560–566.
Dodds, W. J. 1974. Hereditary and acquired hemorrhagic disorders in animals. In T. H. Spaet (ed.), *Progress in Hemostasis and Thrombosis.* Vol. 2. New York: Grune & Stratton, pp., 215–247.
Dodds, W. J. 1975. Further studies of canine von Willebrand's disease. *Blood, 45:* 221–230.

Dodds, W. J. 1976. Inherited bleeding disorders. *Pure-Bred Dogs Amer. Kennel Gazette., 93:* 31–38.

Dodds, W. J., M. A. Packham, H. C. Rowsell, and J. F. Mustard. 1967. Factor VII survival and turnover in dogs. *Amer. J. Physiol., 213:* 36–42.

Donovan, R. H., H. M. Freeman, and C. L. Schepens. 1969. Anomaly of the Collie eye. *J. Amer. Vet. Med. Assoc., 155:* 872–875.

Donovan, R. H., and A. M. Macpherson. 1968. The inheritance of chorioretinal changes and staphyloma in the Collie. *Carnivore Genetics Newsletter,* No. 5, pp. 85–89.

Draper, D. D., J. P. Kluge, and W. J. Miller. 1976. Clinical and pathologic aspects of spinal dysraphism in dogs. *Proc. 20th World Vet. Congr. Thessaloniki, 1975.* Vol. 1. Thessaloniki, Greece. Pp. 134–137.

Dyrendahl, S., and B. Henricson. 1960. Hereditary hyperplastic gingivitis of silver foxes. *Acta Vet. Scand., 1:* 121–139.

Eberhart, G. W. 1959. Epilepsy in the dog. In *9th Gaines Vet. Sympos.* Kankakee, Ill.: Gaines Dog Research Center. Pp. 18–20.

Ewing, G. O. 1969. Familial nonspherocytic hemolytic anemia of Basenji dogs. *J. Amer. Vet. Med. Assoc., 154:* 503–507.

Falco, M. J., J. Barker, and M. E. Wallace. 1974. The genetics of epilepsy in the British Alsatian. *J. Small Anim. Pract., 15:* 685–692.

Field, R. A., C. G. Rickard, and F. B. Hutt. 1946. Hemophilia in a family of dogs. *Cornell Vet., 36:* 285–300.

Finco, D. R., H. J. Kurtz, D. G. Low, and V. Perman. 1970. Familial renal disease in Norwegian Elkhound dogs. *J. Amer. Vet. Med. Assoc., 156:* 747–760.

Fletch, S. M., and P. H. Pinkerton. 1972. An inherited anaemia associated with hereditary chondrodysplasia in the Alaskan Malamute. *Can. Vet. Jour., 13:* 270–271.

Ford, L. 1969. Hereditary aspects of human and canine cyclic neutropenia. *J. Hered., 60:* 293–299.

Fox, M. W. 1963. Inherited inguinal hernia and mid-line defects in the dog. *J. Amer. Vet. Med. Assoc., 143:* 602–604.

Fox, M. W. 1964. Inherited polycystic mononephrosis in the dog. *J. Hered., 55:* 29–30.

Fox, M. W., W. G. Hoag, and J. Strout. 1965. Breed susceptibility, pathogenicity and epidemiology of endemic coliform enteritis in the dog. *Lab. Anim. Care., 15:* 194–200.

François, J. 1961. *Heredity in Ophthalmology.* St. Louis: C. B. Mosby. 731 pp., 629 figs.

Freudiger, U. 1965. Die kongenitale Nierenrindenhypoplasie beim bunten Cocker-Spaniel. *Schweiz. Arch. Tierheilk., 107:* 547–566.

Freudiger, U., V. Schärer, J.-C. Buser, and R. Mühlebach. 1973a. Die Hüftgelenksdysplasie: Bekämpfungsverfahren und Frequenz bei den verschiedenen Rassen. *Schweiz. Arch. Tierheilk., 115:* 69–73.

Freudiger, U., V. Schärer, J.-C. Buser, and R. Mühlebach. 1973b. Die Resultate der Hüftgelenksdysplasie-Bekämpfung beim D. Schäfer in der Zeit von 1965 bis 1972. *Schweiz. Arch. Tierheilk., 115:* 169–173.

Fritz, T. E., L. S. Lombard, S. A. Tyler, and W. P. Norris. 1976. Pathology and familial incidence of orchitis and its relation to thyroiditis in a closed Beagle colony. *Exper. Molec. Pathol., 24:* 142–158.

Fritz, T. E., R. C. Zeman, and M. R. Zelle. 1970. Pathology and familial incidence of thyroiditis in a closed Beagle colony. *Exper. Molec. Pathol., 12:* 14–30.

Gage, E. D. 1975. Incidence of clinical disc disease in the dog. *J. Amer. Anim. Hosp. Assoc., 11:* 135–138.

Green, E. L. 1957. Mutant stocks of cats and dogs offered for research. *J. Hered., 48:* 56–57.
Griffith, J. R., and J. D. Duncan. 1973. Myotonia in the dog: A report of four cases. *Vet. Rec., 93:* 184–188.
Grüneberg, H., and A. J. Lea. 1940. An inherited jaw anomaly in long-haired Dachshunds. *J. Genetics, 39:* 285–296.
Gustavsson, I. 1964. The chromosomes of the dog. *Hereditas, 51:* 187–189 + 2 plates.
Hall, D. E. 1972. *Blood Coagulation and its Disorders in the Dog.* London: Baillière Tindall; Baltimore: Williams and Wilkins. XII + 188 pp.
Hansen, H.-J. 1968. Historical evidence of an unusual deformity in dogs ("Short-Spine Dog"). *J. Small Anim. Pract., 9:* 103–108.
Hare, W. C. D. 1976. Intersexuality in the dog. *Can. Vet. Journ., 17:* 7–15.
Hare, W. C. D., and K. Bovee. 1974. A chromosomal translocation in Miniature Poodles. *Vet. Rec., 95:* 217–218.
Härtl, J. 1938. Die Vererbung des Kryptorchismus beim Hund. *Kleintier u. Pelztier, 14:* 1–37. Cited in *Animal Breed. Abstr., 8:* 278–279, 1940.
Harvey, A. M., and H. N. Christensen. 1964. Uric acid transport system: apparent absence in erythrocytes of the Dalmatian Coach Hound. *Science, 145:* 826–827.
Hayes, H. M. 1974. Congenital umbilical and inguinal hernias in cattle, horses, swine, dogs, and cats: Risk by breed and sex among hospital patients. *Amer. J. Vet. Res., 35:* 839–842.
Hegreberg, G. A., G. A. Padgett, J. R. Gorham, and J. B. Henson. 1969. A connective tissue disease of dogs and mink resembling the Ehlers-Danlos syndrome of man. II: Mode of inheritance. *J. Hered., 60:* 249–254.
Hegreberg, G. A., G. A. Padgett, and J. B. Henson. 1970a. Connective tissue disease of dogs and mink resembling Ehlers-Danlos syndrome of man. III: Histopathologic changes of the skin. *Arch. Pathol., 90:* 159–166.
Hegreberg, G. A., G. A. Padgett, and R. C. Page. 1970b. The Ehlers-Danlos syndrome of dogs and mink. In *Animal Models for Biomedical Research.* Vol. 3. Washington, D.C.: Nat. Acad. Sci., pp. 80–90.
Henricson, B., I. Norberg, and S.-E. Olsson. 1966. On the etiology and pathogenesis of hip dysplasia: A comparative review. *J. Small Anim. Pract., 7:* 673–688.
Heywood, R. 1971. Juvenile cataracts in the Beagle dog. *J. Small Anim. Pract., 12:* 171–177.
Hirth, R. S., and S. W. Nielsen. 1967. A familial canine globoid cell leukodystrophy ("Krabbe type"). *J. Small Anim. Pract., 8:* 569–575.
Hodgman, S. F. J. 1963. Abnormalities and defects in pedigree dogs. 1: An investigation into the existence of abnormalities in pedigree dogs in the British Isles. *J. Small Anim. Pract., 4:* 447–456.
Hodgman, S. F. J., H. B. Parry, W. J. Rasbridge, and J. D. Steel. 1949. Progressive retinal atrophy in dogs. 1: The disease in Irish Setters (Red). *Vet. Rec., 61:* 185–190.
Hofmeyer, C. F. B. 1963. Dermoid sinus in the Ridgeback dog. *J. Small Anim. Pract., 4* (suppl.): 5–8.
Høst, P., and S. Sreinson. 1936. Hereditary cataract in the dog [translated title]. *Norsk. Vet.-Tidsskr., 48:* 244–270. Cited by Burns, M., and M. Fraser, 1966 (Philadelphia: Lippincott).
Hutt, F. B. 1958. *Genetic Resistance to Disease in Domestic Animals.* Ithaca, N.Y.: Cornell Univ. Press. XIII + 198 pp.
Hutt, F. B. 1964. *Animal Genetics.* New York: Ronald Press. XIV + 546 pp.
Hutt, F. B. 1967. Genetic selection to reduce the incidence of hip dysplasia in dogs. *J. Amer. Vet. Med. Assoc., 151:* 1041–1048.

Hutt, F. B., and A. de Lahunta. 1971. A lethal glossopharyngeal defect in the dog. *J. Hered., 62:* 291–293.
Hutt, F. B., and M. C. Nesheim. 1968. Polygenic variation in the utilization of arginine and lysine by the chick. *Canad. J. Genet. Cytol., 10:* 564–574.
Hutt, F. B., C. G. Rickard, and R. A. Field. 1948. Sex-linked hemophilia in dogs. *J. Hered., 39:* 2–9.
Johnston, D. E., and B. Cox. 1970. The incidence in purebred dogs in Australia of abnormalities that may be inherited. *Austral. Vet. J., 46:* 465–474.
Karbe, E., and Schiefer, B. 1967. Familial amaurotic idiocy in male German Shorthair Pointers. *Path. Vet., 4:* 223–232.
Keeler, C. E. 1940. The inheritance of predisposition to renal calculi in the Dalmatian. *J. Amer. Vet. Med. Assoc., 96:* 507–510.
Klarenbeek, A., S. Koopmans, and J. Winsser. 1942. Eenaanbalsgewijs optredende stoornis in de regulatie van de spiertonus, waargenomen bij Schotsche Terriers. *Tijdschr. Diergeneesk., 69:* 14–21.
Koch, S. A. 1972. Cataracts in Old English Sheepdogs. *J. Amer. Vet. Med. Assoc., 160:* 299–301.
Koppang, N. 1970. Neuronal ceroid-lipofuscinosis in English Setters. *J. Small Anim. Pract., 10:* 639–644.
Krook, L. 1957. The pathology of renal cortical hypoplasia in the dog. *Nord. Vet. Med., 9:* 161–176.
Ladrat, J., P.-C. Blin, and J.-J. Lauvergne. 1969. Ectromélie bithoracique héréditaire chez le chien. *Ann. Génét. Sél. anim., 1:* 119–130.
Lau, R. E. 1977. Inherited premature closure of the distal ulnar physis. *J. Amer. Anim. Hosp. Assoc., 13:* 609–612.
Letard, É. 1930. Le Mendélisme expérimental. Expériences sur hérédité mendelienne du caractère "peau nue" dans l'espèce chien. *Rév. Vét. et J. Méd. Vét. (Toulouse), 82:* 553–570.
Little, C. C. 1948. Genetics in Cocker Spaniels: Observations on heredity and on physiology of reproduction in American Cocker Spaniels. *J. Hered., 39:* 181–185.
Little, C. C. 1957. *The Inheritance of Coat Color in Dogs.* Ithaca, N.Y.: Cornell University Press. Reprinted by Howell Book House (New York). XIII + 194 pp.
Loeffler, K., and H. Meyer. 1961. Erbliche Patellarluxation bei Toy-Spaniels. *Deutsche Tierärztl. Wochenschrift, 68:* 619–622.
Lund, J. E., G. A. Padgett, and J. R. Gorham. 1970. Additional evidence on the inheritance of cyclic neutropenia in the dog. *J. Hered., 61:* 46–49.
Ma, N. S. F., and C. E. Gilmore. 1971. Chromosomal abnormality in a phenotypically and clinically normal dog. *Cytogenetics, 10:* 254–259.
McClearn, G. E., and J. C. DeFries. 1973. *Introduction to Behavioral Genetics.* San Francisco: W. H. Freeman. IX + 349 pp.
McGrath, J. T. 1965. Spinal dysraphism in the dog. *Pathologia Veterinaria, 2* (Suppl.): 1–36 + 46 figs.
Mann, G. E., and J. Stratton. 1966. Dermoid sinus in the Rhodesian Ridgeback. *J. Small Anim. Pract., 7:* 631–642.
Martin, C. L., and H. W. Leipold. 1974. Aphakia and multiple ocular defects in Saint Bernard puppies. *Vet. Med. Small Anim. Clinician, 69:* 448–453.
Merkens, J. 1938. Haemophilie bij honden. *Ned.-ind. Bl. Diergeneesk., 50:* 149–151.
Meyers, K. M., J. E. Lund, G. A. Padgett, and W. M. Dickson. 1969. Hyperkinetic episodes in Scottish Terrier dogs. *J. Amer. Vet. Med. Assoc., 155:* 129–133.
Meyers, K. M., G. A. Padgett, and W. M. Dickson. 1970. The genetic basis of a kinetic disorder of Scottish Terrier dogs. *J. Hered., 61:* 189–192.

Minouchi, O. 1928. The spermatogenesis of the dog, with special reference to meiosis. *Jap. J. Zool., 1:* 255–268.

Mitchell, A. L. 1935. Dominant dilution and other color factors in Collie dogs. *J. Hered., 26:* 424–430.

Moore, L. A., and P. J. Schaible. 1936. Inheritance of umbilical hernia in rats. *J. Hered., 27:* 272–280.

Moore, W., Jr., and P. D. Lambert. 1963. The chromosomes of the Beagle dog. *J. Hered., 54:* 273–276.

Morgan, J. P., G. Ljunggren, and R. Read. 1967. Spondylosis deformans (vertebral osteophytosis) in the dog. *J. Small Anim. Pract., 8:* 57–66.

Mühlebach, R., and U. Freudiger. 1973. Röntgenologische Untersuchungen über die Erkrankungsformen der Spondylose beim Deutschen Boxer. *Schweiz. Arch. Tierheilk., 115:* 539–558.

Musser, E., and W. R. Graham. 1968. Familial occurrence of thyroiditis in purebred Beagles. *Lab. Animal Care, 18:* 58–68.

Mustard, J. F., D. Secord, T. D. Hoeksema, H. G. Downie, and H. C. Rowsell. 1962. Canine factor VII deficiency. *Brit. J. Haematology, 8:* 43–47.

Needler, A. W. H. and R. R. Logie. 1947. Serious mortalities in Prince Edward Island oysters caused by a contagious disease. *Trans. Roy. Soc. Canada.,* 3rd Ser., Sec. V, *41:* 73–89.

Oleson, H. P., O. A. Jensen, and M. S. Norn. 1974. Congenital hereditary cataract in Cocker Spaniels. *J. Small Anim. Pract., 15:* 741–750.

Olsson, S.-E. 1963. Höftledsdysplasin på Tillbakagång. *Hundsport., 11:* 16–19.

Osborne, C. A., D. H. Clifford, and C. Jessen. 1967. Hereditary esophageal achalasia in dogs. *J. Amer. Vet. Med. Assoc., 151:* 572–581.

Palmer, A. C., J. E. Payne, and M. E. Wallace. 1973. Hereditary quadriplegia and amblyopia in the Irish Setter. *J. Small Anim. Pract., 14:* 343–352.

Patel, V., N. Koppang, B. Patel, and W. Zeman. 1974. *p*-Phenylenediamine-mediated peroxidase deficiency in English Setters with neuronal ceroid-lipofuscinosis. *Lab. Invest., 30:* 366–368.

Patterson, D. F. 1968. Epidemiologic and genetic studies of congenital heart disease in the dog. *Circulation Res., 23:* 171–202.

Patterson, D. F. 1977. A catalogue of genetic disorders of the dog. In R. W. Kirk (ed.), *Current Veterinary Therapy.* Vol. 6. Philadelphia: W. B. Sanders. Pp. 73–89.

Patterson, D. F., W. Medway, H. Luginbühl, and S. Chacko. 1967. Congenital hereditary lymphoedema in the dog. Part 1: Clinical and genetic studies. *J. Med. Genetics, 4:* 145–152.

Patterson, D. F., and R. L. Pyle. 1971. Genetic aspects of congenital heart disease in the dog. Paper presented at 21st Gaines Vet. Symposium, October 20, 1971, Ames, Iowa, pp. 20–28.

Patterson, D. F., R. L. Pyle, J. W. Buchanan, E. Trautvetter, and D. A. Abt. 1971. Hereditary patent ductus arteriosus and its sequelae in the dog. *Circulation Res., 29:* 1–13.

Patterson, D. F., R. L. Pyle, L. V. Mierop, J. Melbin, and M. Olson. 1974. Hereditary defects of the conotruncal septum in Keeshond dogs: pathologic and genetic studies. *Amer. J. Cardiol., 34:* 187–205.

Pfaffenberger, C. J. 1963. *The New Knowledge of Dog Behavior.* New York: Howell. 206 pp.

Phillips, J. McI. 1945. "Pig-jaw" in Cocker Spaniels. *J. Hered., 36:* 177–181.

Phillips, J. M., and T. M. Felton. 1939. Hereditary umbilical hernia in dogs. *J. Hered., 30:* 433–435.

Pick, J. R., R. A. Goyer, J. B. Graham, and J. H. Renwick. 1967. Subluxation of the carpus in dogs. *Lab. Investigation, 17:* 243–248.

Pinkerton, P. H., S. M. Fletch, P. J. Brueckner, and D. R. Miller. 1974. Hereditary stomatocytosis with hemolytic anemia in the dog. *Blood, 44:* 557–567.

Pirchner, F. 1969. *Population Genetics in Animal Breeding.* San Francisco: W. H. Freeman. XI + 274 pp.

Pobisch, R., V. Geres, and E. Arbesser. 1972. Elbogelenksdysplasie beim Hund. *Wiener Tierärztl. Monatsschrift, 59:* 297–307.

Prasse, K. W., D. Crouser, E. Beutler, M. Walker, and W. D. Schall. 1975. Pyruvate kinase deficiency anemia with terminal myelofibrosis and osteosclerosis in a Beagle. *J. Amer. Vet. Med. Assoc., 166:* 1170–1175.

Priester, W. A. 1972. Sex, size, and breed as risk factors in canine patellar dislocation. *J. Amer. Vet. Med. Assoc., 160:* 740–742.

Priester, W. A. 1974. Canine progressive retinal atrophy: Occurrence by age, breed and sex. *Amer. J. Vet. Res., 35:* 571–574.

Rasmussen, P. G. 1972. Multiple epiphyseal dysplasia in a litter of Beagle puppies. *J. Small Anim. Pract., 12:* 91–96.

Read, D. H., D. D. Harrington, T. W. Keenan, and E. J. Hinsman. 1976. Neuronal-visceral GM_1 gangliosidosis in a dog with β-galactosidase deficiency. *Science, 194:* 442–445.

Rehfeld, C. E. 1970. Definition of relationships in a closed Beagle colony. *Amer. J. Vet. Res., 31:* 723–731.

Riser, W. H., D. Cohen, S. Lindquist, J. Mansson, and S. Chen. 1964. Influence of early rapid growth and weight gain on hip dysplasia in the German Shepherd Dog. *J. Amer. Vet. Med. Assoc., 145:* 661–668.

Riser, W. H., and J. F. Shirer. 1967. Correlation between canine hip dysplasia and pelvic muscle mass: A study of 95 dogs. *Amer. J. Vet. Res., 28:* 769–777.

Roberts, S. R. 1967. Color dilution and hereditary defects in Collie dogs. *Amer. J. Ophthalmol., 63:* 1762–1772.

Roberts, S. R., and S. I. Bistner. 1968. Persistent pupillary membrane in Basenji dogs. *J. Amer. Vet. Med. Assoc., 153:* 533–542.

Robinson, R. 1972. Catalogue and bibliography of canine genetic anomalies. (2nd ed.). 52 pp. mimeo. CHART, West Wickham, Kent, England.

Rowsell, H. C. 1963. Hemorrhagic disorders in dogs: Their recognition, treatment and importance. In *The Newer Knowledge about Dogs: 12th Gaines Veterinary Symposium.* Symposium held January 23, 1963, East Lansing, Mich. Pp. 9–16.

Rowsell, H. C., H. G. Downie, J. F. Mustard, J. E. Leeson, and J. A. Archibald. 1960. A disorder resembling hemophilia B (Christmas disease) in dogs. *J. Amer. Vet. Med. Assoc., 137:* 247–250.

Rubin, L. F. 1963. Hereditary retinal detachment in Bedlington Terriers. *Small Anim. Clinician, 3:* 387–389.

Rubin, L. F. 1968. Heredity of retinal dysplasia in Bedlington Terriers. *J. Amer. Vet. Med. Assoc., 152:* 260–262.

Rubin, L. F. 1969. Comments—Collie eye anomaly. *J. Amer. Vet. Med. Assoc., 155:* 865–866.

Rubin, L. F., T. K. R. Bourns, and L. H. Lord. 1967. Hemeralopia in dogs: Heredity of hemeralopia in Alaskan Malamutes. *Amer. J. Vet. Res., 28:* 355–357.

Rubin, L. F., and R. D. Flowers. 1972. Inherited cataract in a family of Standard Poodles. *J. Amer. Vet. Med. Assoc., 161:* 207–208.

Rubin, L. F., S. A. Koch, and R. J. Huber. 1969. Hereditary cataracts in Miniature Schnauzers. *J. Amer. Vet. Med. Assoc., 154:* 1456–1458.

Šanda, A., and J. Křiženecký. 1965. Genetic basis of necrosis of digits in Shortcoated Setters. *Sborník Vysoké Školy Zemědělské V Brně, B: Spisy Fakulty Veterinární, 13:* 281–296. (In English.)

Sanda, A., and L. Pivnik. 1964. Die Zehennekrose bei kurzhaarigen Vorstehhunden. *Die Kleintierpraxis, 9:* 76–83.

Sandefeldt, E., J. F. Cummings, A. de Lahunta, G. Björk, and L. Krook. 1973. Hereditary neuronal abiotrophy in the Swedish Lapland dog. *Cornell Vet., 63:* 1–71.

Schnelle, G. B. 1973. The present status and outlook on canine hip dysplasia. *Gaines Dog Res. Progress,* Spring issue, pp. 1, 6–7.

Scott, J. P., and J. L. Fuller. 1965. *Genetics and the Social Behavior of the Dog.* Chicago and London: University of Chicago Press. XVIII + 468 pp.

Searcy, G. P., D. R. Miller, and J. B. Tasker. 1971. Congenital hemolytic anemia in the Basenji dog due to erythrocyte pyruvate kinase deficiency. *Canad. J. Comp. Med., 35:* 67–70.

Seidenberg, L., and C. D. Knecht. 1971. Ectopic ureter in the dog. *J. Amer. Vet. Met. Assoc., 159:* 876–877.

Selden, J. R., P. S. Moorhead, M. L. Oehlert, and D. F. Patterson. 1975. The Giemsa banding pattern of the canine karyotype. *Cytogenet. Cell Genet., 15:* 380–387.

Selden, J. R., S. S. Wachtel, G. C. Koo, M. E. Haskins, and D. F. Patterson. 1978. Genetic basis of XX male syndrome and XX true hermaphroditism: evidence in the dog. *Science,* 201: 644–646.

Selmanowitz, V. J., K. M. Kramer, and N. Orentreich. 1970. Congenital ectodermal defect in Miniature Poodles. *J. Hered., 61:* 196–199.

Shive, R. J., W. C. D. Hare, and D. F. Patterson. 1965. Chromosome studies in dogs with congenital heart defects. *Cytogenet., 4:* 340–348.

Slappendel, R. J. 1975. Hemophilia A and hemophilia B in a family of French Bulldogs. *Tijdschr. Diergeneesk., 100:* 1075–1088.

Sorsby, A., and J. B. Davey. 1954. Ocular associations of dappling (or merling) in the coat colour of dogs. 1: Clinical and genetical data. *J. Genet., 52:* 425–440.

Spurling, N. W., L. K. Burton, R. Peacock, and T. Pilling. 1972. Hereditary factor VII deficiency in the Beagle. *Brit. J. Haematol., 23:* 59–67.

Spurling, N. W., L. K. Burton, and T. Pilling. 1974a. Canine factor VII deficiency: Experience with a modified thrombotest method in distinguishing between the genotypes. *Research in Vet. Science, 16:* 228–239.

Spurling, N. W., R. Peacock, and T. Pilling. 1974b. The clinical aspects of canine factor VII deficiency including some case histories. *J. Small Anim. Pract., 15:* 229–239.

Stevens, R. W. C., and M. E. Townsley. 1970. Canine serum transferrins. *J. Hered., 61:* 71–73.

Stockard, C. R. 1936. An hereditary lethal for localized motor and preganglionic neurones with a resulting paralysis in the dog. *Amer. J. Anat., 59:* 1–53.

Stockard, C. R. 1941. The genetic and endocrinic basis for differences in form and behavior as elucidated by studies of contrasted pure-line dog breeds and their hybrids. *Amer. Anatom. Memoirs, 19.* XX + 1–775. Philadelphia: Wistar Inst. Anat. Biol.

Stockard, C. R., and A. L. Johnson. 1941. The contrasted patterns and modifications of head types and forms in the pure breeds of dogs and their hybrids as the results of genetic and endocrinic reactions. *Amer. Anat. Memoirs, 19.* Sec. III, pp. 149–383. Philadelphia: Wistar Inst. Anat. Biol.

Subden, R. E., S. M. Fletch, M. A. Dmart, and R. G. Brown. 1972. Genetics of the Alaskan Malmute [*sic*] chondrodysplasia syndrome. *J. Hered., 63:* 149–152.

Swisher, S. N., and L. E. Young. 1961. The blood grouping systems of dogs. *Physiol. Rev., 41:* 495–520.

Tasker, J. B., G. A. Severin, S. Young, and E. L. Gillette. 1969. Familial anemia in the Basenji dog. *J. Amer. Vet. Med. Assoc., 154:* 158–165.
Templeton, J. W., A. P. Stewart, and W. S. Fletcher. 1977. Coat color genetics in the Labrador Retriever. *J. Hered., 68:* 134–136.
Thomsett, L. R. 1961. Congenital hypotrichia in the dog. *Vet. Rec., 73:* 915–918.
Trimble, H. C., and C. E. Keeler. 1938. The inheritance of "high uric acid excretion" in dogs. *J. Hered., 29:* 280–289.
Van Der Velden, N. A. 1968. Fits in Tervueren Shepherd dogs: A presumed hereditary trait. *J. Small Anim. Pract., 9:* 63–70.
Vriesendorp, H. M., B. D. Hartog, B. M. J. Smid-Mercx, and D. L. Westbroek. 1973. Immunogenetic markers in canine paternity cases. *J. Small Anim. Pract., 15:* 693–699.
Vriesendorp, H. M., et al. [25 contributors]. 1976. Joint report of the second international workshop on canine immunogenetics. *Transplantation Proc., 8:* 289–314.
Waardenburg, P. J. 1932. Das menschliche Auge und seine Erbanlagen. [The human eye from the genetic standpoint]. *Bibliographia Genetica.* Deel. VII [Vol. 7]. The Hague: Martinus Nijhoff. XVI+ 631 pp.
Warren, D. C. 1927. Coat color inheritance in Greyhounds. *J. Hered., 18:* 512–522.
Warwick, B. L. 1931. Breeding experiments with sheep and swine. Ohio Agric. Exper. Sta. Bull. 480. 37 pp.
Warwick, B. L. 1932. Probability tables for Mendelian ratios with small numbers. Texas Agric. Exper. Sta. Bull. 463. 28 pp.
Weber, W. 1959. Über die Vererbung der medianen Nasenspalte beim Hund. *Schweiz. Arch. Tierheilk., 101:* 378–381.
Wegner, W. 1975. *Kleine Kynologie für Tierärzte und andere Tierfreunde.* Konstanz: Terra-Verlag. 224 pp.
Wentinck, G. H., W. Hartman, and J. P. Koeman. 1974. Three cases of myotonia in a family of Chows. *Tijdschr. Diergeneesk., 99:* 729–731.
Wentink, G. H., J. S. van der Linde-Sipman, A. E. F. H. Meijer, H. A. C. Kamphuisen, C. J. A. H. V. van Vorstenbosch, W. Hartman, and H. J. Hendricks. 1972. Myopathy with a possible recessive X-linked inheritance in a litter of Irish Terriers. *Vet. Pathol., 9:* 328–349.
White, E. G., R. J. Treacher, and P. Porter. 1961. Urinary calculi in the dog. 1: Incidence and chemical composition. *J. Comp. Pathol. Therap., 71:* 201–216.
Whitney, L. F. 1939. The sex ratio of dogs maintained under similar conditions. *J. Hered., 30:* 388–389.
Whitney, L. F. 1948. *How to Breed Dogs.* (Rev. ed.) New York: Orange Judd.
Whitney, L. F. 1972. *How to Breed Dogs.* (Rev. ed.). New York: Howell. 384 pp.
Wilkinson, J. S. 1960. Spontaneous diabetes mellitus. *Vet. Rec., 72:* 548–558.
Willeberg, P., K. W. Kastrup, and E. Andresen. 1975. Pituitary dwarfism in German Shepherd Dogs: Studies on somatomedin activity. *Nord. Vet. Med., 27:* 448–454.
Willis, M. B. 1963. Abnormalities and defects in pedigree dogs. 5: Cryptorchidism. *J. Small Anim. Pract., 4:* 469–474.
Winge, O. 1950. *Inheritance in Dogs.* Ithaca, N.Y.: Cornell University Press. IX + 153 pp.
Wriedt, C. 1925. Letale Faktoren [Todbringende Vererbungsfaktoren]. *Ztschr. Tierzüchtung u. Züchtungsbiol., 3:* 223–230.
Wright, S. 1931. Evolution in Mendelian populations. *Genetics, 16:* 97–159.
Yakely, W. L. 1972. Collie eye anomaly: Decreased prevalence through selective breeding. *J. Amer. Vet. Med. Assoc., 161:* 1103–1107.

Yakely, W. L. 1975. Cataracts in the American Cocker Spaniel. *Proc. Amer. Coll. Vet. Ophthalmologists.* Conference held September 21, 1975, Dallas, pp. 27–34.

Yakely, W. L., G. A. Hegreberg, and G. A. Padgett. 1971. Familial cataracts in the American Cocker Spaniel. *J. Amer. Anim. Hosp. Assoc., 7:* 127–135.

Yakely, W. L., M. Wyman, E. F. Donovan, and N. S. Fechheimer. 1968. Genetic transmission of an ocular fundus anomaly in Collies. *J. Amer. Vet. Med. Assoc., 152:* 457–461.

Glossary

Definitions of genetical terms used in this book and of a few others that readers may encounter elsewhere.

Allele Either of two genes that occupy the same locus in homologous chromosomes, or of the two contrasting characters induced by such genes. For example, the gene *B* causing black coat color is the dominant allele of *b*, which induces brown.

Aneuploid An adjective, or noun, applied to an organism that has more or less than (1) the number of chromosomes normal for the species and sex, or (2) some multiple of that number. See *monosomic* and *trisomic.*

Autosomal Pertaining to an autosome, or to characters induced by autosomes, in contradistinction to sex-linked characters.

Autosome Any chromosome other than a sex chromosome.

Backcross The mating of an F_1 hybrid or of an equivalent heterozygote to either of the two parental varieties, types, or genotypes that produced the hybrid.

Bimodal An adjective applied to a distribution of frequencies that shows two classes as having greater numbers than others.

Breeding test A mating, usually between one parent of unknown genotype and one of known genotype, from which the progeny will reveal the genotype that is unknown.

Carrier An organism heterozygous for some recessive character.

Character A convenient term used to designate any structure, trait, or function of an organism, whether it be hereditary or acquired.

Centromere The point in a chromosome at which the spindle-fibre (seen during cell division) is attached.

Checkerboard A square or rectangle showing all possible combinations of male and female gametes in some particular cross. Also called a Punnett square.

Chromosomes Literally, colored bodies. When stained with basic dyes, they are visible under the microscope as rods, loops, or dots in dividing cells. They carry the genes, which are arranged in linear order.

Chromosome map A diagram showing the linkage relationships and relative distances apart of genes proven by linkage tests to be in one and the same chromosome. Bigger maps show such diagrams for several chromosomes.

Codominant An adjective applicable to a pair of alleles both of which exert approximately equal effects on the phenotype, so that the heterozygote is intermediate between the two homozygotes.

Complementary genes Genes which in combination cause an effect different from that of either gene by itself.

Congenital Present at birth (perhaps unseen) but not necessarily of genetic origin.

Continuous variation That in which there are many intergrading forms (none sharply divided from the others) between the two extremes of some polygenic character.

Crossing-over An interchange of parts of homologous chromosomes (and hence of genes) during their division in the formation of germ cells when the number of chromosomes is reduced.

Crossover An individual that, because of crossing over during formation of one (or both) of the gametes from which it arose, carries the alleles of two or more linked genes in a combination different from that in which they were transmitted to one (or both) of its parents. To illustrate, in Figure 4-5, if there had been any crossing over during formation of gametes in the five dihybrid bitches that were

$$\frac{H\ f}{h\ F}$$

that could have resulted in male crossovers of the genotype $h\ f$ (flat-footed, and hemophilic), or $H\ F$ (with neither recessive allele). See Chapter 4.

Coupling phase That type of association of two linked pairs of genes in which the chromosome carrying them has either both dominant alleles or both recessive alleles; for example, $H\ F$ or $h\ f$. See also *repulsion phase.*

Cytology The study of cells, their structure, contents, and functions.

Cytoplasm The protoplasm of the cell exclusive of the nucleus.

Diallel crosses The mating of two or more animals (at different times) to two or more animals of the opposite sex in order to evaluate better the genotypes of all concerned with respect to some polygenic character; one kind of progeny-test.

Dihybrid Heterozygous in two known pairs of alleles.

Diploid Having two sets of chromosomes, as in somatic cells of animals, in contradistinction to the haploid state of the germ cells, which have only a single set of chromosomes.

Discontinuous variation That in which certain types are clearly distinct from others.

DNA An abbreviation for deoxyribonucleic acid (of which the gene is formed).

Dominant Referring to genes or characters that are manifested by organisms heterozygous for them, in contradistinction to recessive genes or characters, which are revealed only in homozygotes.

Duplicate genes Genes at different loci that have equal effects on one character.

Dysgenic Conducive to the accumulation of undesirable genes in the germ plasm of a species, breed, or population and hence to the weakening of the racial fitness of future generations; the opposite of *eugenic.*

Egg A female reproductive cell.

Epistasis The masking of the action of one gene (or pair of alleles) by another at a different locus.

Epistatic An adjective describing a gene or character which obscures others that are not its alleles. For example, Winge postulated an epistatic red in Irish Setters that would otherwise be black.

Eugenic Conducive to the improvement of "the racial qualities of future generations, either physically or mentally" (Galton).

Expressivity Degree to which a character is manifested by those that show any trace of it.

F_1 generation The first filial generation; the first generation from some specific mating.

F_2 generation The second filial generation from some specific mating. It is produced by mating individuals of the F_1 generation *inter se.*

Factor An earlier name for what is now called a *gene.* Also used in other disciplines, as for factors I to XIII, which influence blood clotting.

Gamete A reproductive cell, whether an egg or a sperm.

Gene A unit of inheritance, now known to be composed of DNA.

General combining ability Ability of one strain to yield good results in most crosses with other strains. See also *Specific combining ability.*

Genetics The science that deals with heredity and variation, seeking to elucidate the principles underlying the former and causes for the latter.

Genetic junk Jargon (in the geneticist's lexicon) for designating collectively genes that are undesirable.

Genotype The genetic constitution of an organism, including genes without visible effects as well as those revealed by the phenotype. It may refer to all the genes or to a single pair of alleles.

Germ cell A reproductive cell, or one capable of giving rise to gametes, in contradistinction to somatic cells (body cells), which cannot do so.

Germ plasm The sum total of inheritance that is passed along from one generation to the next.

Gynandromorph An animal part of which is male and part female.

Haploid Single; referring to the reduced number of chromosomes as found in gametes.

Hemizygous Referring to the unpaired state of sex-linked genes in individuals of the heterogametic sex. For example, a male dog can be neither homozygous nor heterozygous for hemophilia A. It is hemizygous.

Heritability Usually designated as h^2, and purported to measure, on a scale between 0.1 and 1.0, how much of the variation in a polygenic character is of genetic origin.

Heterogametic Producing gametes of two kinds with respect to sex chromosomes, the difference being that one kind induces (at fertilization) a male zygote, and the other a female.

Heterosis Hybrid vigor.

Heterozygous (n., heterozygote) Carrying both the dominant and recessive genes of a pair of alleles, or two different genes of a series of multiple alleles.

Homogametic Producing gametes that are all equipotential with respect to the sex of the zygote to which they give rise on fertilization.

Homologous chromosomes Those that occur in somatic cells in matching pairs, both being alike in size and form, and differing from each other (with respect to genes) less than they differ from other pairs. One member of a pair is normally inherited from each parent.

Homozygous (n., homozygote) Carrying two of either the dominant or recessive genes of a pair of alleles, or carrying two identical genes of a series of multiple alleles.

Hybrid In general biological usage, this term refers to the offspring of a cross between different species or varieties. Geneticists apply it also to organisms heterozygous with respect to one or more pairs of genes.

Hybrid vigor The extra vigor, exceeding that of their parent stocks, frequently shown by the hybrids from crosses of different species, breeds, strains, or inbred lines. It may be expressed as more rapid growth, larger size, greater productivity, or otherwise.

Hypostatic A character masked by some epistatic character. For example, animals with white coats or plumage can carry genes for various patterns, but those patterns are not visible. They are hypostatic.

Inbreeding The mating of relatives.

Inbreeding, coefficient of A measure of the extent to which inbreeding has reduced heterozygosity in comparison with that prevailing in similar animals not inbred.

Inter se Among themselves.

Karyotype An illustration showing the chromosomes of a single cell arranged in matching pairs, usually in descending order of size.

Lethal gene One that causes premature death of the organism homozygous or hemizygous for it. That death may occur at any period between fertilization of the egg and the normal life span of the species, but, in a narrower sense, the term is sometimes restricted to genes causing death before birth, at birth, or soon thereafter.

Line breeding The mating of later generations back to some ancestor or its descendants.

Linkage The association of two or more genes or characters in inheritance because the genes are on the same chromosome.

Linkage group A group of genes or characters each of which has been shown to be linked with other members of the same group. The number of linkage groups in a species cannot exceed the number of pairs of chromosomes that is normal for the species.

Locus (pl., loci) The position of a gene in a chromosome or linkage group, usually stated in terms of its distance from other genes in that chromosome.

Mass selection Selection of breeding stock based on the appearance or performance of individuals; that is, on the phenotype.

Matroclinous Resembling the mother.

Mendelian ratios Those expected from the segregation of only a few characters (usually 1 to 3) from various matings in which the characters (either separately or in combinations) are fully manifested; for example, 3 : 1, 1 : 1, or (in F_2 from dihybrids) 9 : 3 : 3 : 1.

Modifier (or modifying gene) A gene, usually with slight effects, that influences the expression of some genetic character.

Monogenic Caused by a single pair of genes.

Monosomic Having one fewer than the number of chromosomes normal for the species. See also *Trisomic.*

Morphological character A genetic variation in form or structure, in contradistinction to physiological characters, which are variations in function.

Multifactorial Inheritance based on an undetermined number of genes. Same as *Polygenic.*

Multiple alleles A series of three or more genes any of which may occupy a single locus on a chromosome. All of them influence the same character, but in differing degrees.

Multiple factors An undetermined number of genes that together influence the expression of one genetic character.

Mutant Used as an adjective to describe a gene that has undergone a mutation, or as a noun to designate an individual that shows the result of that mutation.

Mutation A sudden change in a gene or in the chromosomes, resulting in a new variation that is hereditary. It may refer either to the invisible change in the cells or to the resultant visible change in the organism.

Normal distribution Usually applied to a frequency distribution of some polygenic character in which the number of individuals in classes above the mean is approximately the same as the number below it.

Normal overlaps Jargon, perhaps, but referring to individuals indistinguishable from normal ones even though their genotype usually causes some expression of a genetic character. They occur in populations showing continuous intergradation between some hereditary character and the normal type.

Nucleus A small body within the cell that stains deeply with basic dyes, contains the chromosomes, and reproduces itself in cell division.

Outcross Mating of a stock that is somewhat inbred to unrelated individuals.

Ovum An egg at the stage when it is ready for fertilization.

Patroclinous Resembling the sire.

Penetrance The extent to which a character is expressed in a group of individuals homozygous for it; measured as the proportion (percent) of such individuals that show the character.

Phenotype The genetic nature of an organism that is revealed by visible characters or measurable performance, in contradistinction to the genotype, which may not be evident without a breeding test.

P_1 generation Parental generation.

Pleiotropic An effect of a gene on form or function in ways other than the most visible one by which the phenotype was originally recognized; for example, cyclic neutropenia in the gray Collie; stomatocytosis in the dwarf Malamutes.

Polygenic Dependent on the combined action of an undetermined number of genes.

Polymorphism Occurring in several different forms.

Polyploid Referring to cells (or an organism) having three or more complete sets of homologous chromosomes; i.e., at least one set in excess of the normal diploid number.

Proband (or propositus) The individual (in a pedigree) from which the investigation was begun and the pedigree compiled.

Progeny-test Evaluation of the genotype of an animal by the kinds of offspring that it produces.

Protoplasm The active, living substance of the cell, including the nucleus and the cytoplasm.

Proven sire One with enough progeny that have been measured (by appearance or performance) to reveal the inheritance transmitted to them, whether good or bad, and hence to give some idea of the genotype of their sire.

Quantitative character An inherited character showing continuous intergradation between the extremes of its expression; a polygenic character.

Recessive An adjective, frequently used as a noun, referring to a gene or character that is expressed only by homozygotes; that gene of a pair of alleles that does not show in heterozygotes.

Reciprocal crosses Crosses made both ways between strains or breeds; for example, ♂ A × ♀ B and ♀ A × ♂ B.

Repulsion phase That type of association of two linked pairs of genes in which the chromosomes carrying them have the dominant allele of one pair and the recessive allele of the other; for example, *H f* or *h F*. See also *Coupling phase.*

Reversion The occurrence of an individual that differs in some respects from its parents but resembles a grandparent or some ancestor more remote; sometimes used with reference to domestic animals that resemble the wild-type ancestor from which the domestic forms have arisen.

Segregation The separation (during reduction of the chromosomes) of allelic genes and the random recombination (at fertilization) of the two kinds of gametes thus produced, with the result that different types appear in the progeny in typical Mendelian ratios.

Self In genetic parlance, uniform (solid) color.

Selection pressure The degree to which the animals chosen to beget the next generation excel the average of their own generation.

Semi-lethal Lethal to some of the animals having the genotype for its expression; does *not* mean that it leaves any of them half dead.

Sex chromosomes Those that exert a preponderant influence on the determination of sex. In animals, the homogametic sex has two homologous sex chromosomes, and the heterogametic sex has either only one of these or one with a dissimilar partner.

Sex dimorphism A difference between the sexes in size, structure, color, or some other attribute.

Sex-limited Manifested only in one sex.

Sex-linked An adjective applied to a gene carried in the kind of sex chromosome that is paired in the homogametic sex, or to a character induced by such a gene.

Sex ratios Expressed either as ♂♂ per 100 ♀♀ or as the proportion (percent) of males in a population; at conception, *primary* sex ratio; at birth, *secondary* sex ratio; at later specified ages, *tertiary* sex ratio.

Siblings Brothers and sisters.

Sib-test Estimation of the genotype for an individual from the appearance or performance of its siblings.

Simple In genetic parlance, this adjective has a special connotation when it is applied to characters dependent for their expression on a single pair of alleles.

Somatic Referring to all cells and tissues other than the germ cells.

Specific combining ability Applied to a strain that gives good results in crosses with some strains, but not with all. See also *General combining ability*.

Sport An organism differing sharply from the normal for the breed or variety; a mutation.

Subvital An adjective describing genes or characters that reduce the biological efficiency of an organism.

Symbol One or two letters (sometimes more), or a combination of letters and numbers, used to designate a gene or character so that its name need not be written out in full.

Test-cross Same as backcross, or mating of suspected carrier to known carrier.

Trisomic Having one more than the number of chromosomes normal for the species. See also *Monosomic*.

Unifactorial An adjective applied to a character that is dependent for its expression upon the action of a single pair of genes. If dominant, the character is shown by the heterozygote; if recessive, it is revealed only by homozygotes.

Wild type The normal phenotype of a species as it occurs in nature, or (in domestic animals) as it is presumed to have been prior to domestication. It is a convenient norm by comparisons with which mutations may be designated as dominant or recessive to the wild type.

W chromosome For species having heterogametic females (as in birds and *Lepidoptera*), this term is often used to designate the smaller of the two sex chromosomes. It occurs only in the females and is the counterpart of the Y chromosome of mammals.

X chromosome The designation used for that sex chromosome that, in species having heterogametic males (as do mammals), is paired in females, but occurs singly in males. See also *Z chromosome*.

Y chromosome The sex chromosome that normally occurs singly in heterogametic males and does not occur in females of the same species.

Z chromosome A designation often used for the sex chromosome of species having heterogametic females that is paired in males but occurs singly in females. Thus ZZ-ZW in birds are counterparts of the sex chromosomes XX-XY in mammals. The only reason for using Z and W is to indicate that the female is heterogametic.

Zygote The cell formed by union of male and female gametes; the fertilized egg.

Appendix I

Lethal and Semi-lethal Characters in Dogs

Character	Symbol[a]	Lethal to	Age at Death	Discussed in Chapter
Bird-tongue	*bt*	*bt bt*	1–3 days	1, 6
Extreme hairlessness	*N*	*NN*	mostly before birth	1, 2, 8
Hemophilia A	*h8*	some *h8* Y	various	4
Hemophilia B	*h9*	some *h9* Y	various	4
Necrosis of toes	—	all	4–8 months	8
Thoracic ectromelia	—	nearly all	in first week	7
Ataxia	*at*	*at at*	in one breed: about 20 weeks	9
Neuronal abiotrophy	*na*	*na na*	7 weeks to 18 months	9
Sex-linked myopathy	*my*	*my* Y	several weeks	9
Globoid-cell leukodystrophy	*gc*	*gc gc*	about 6 months	9
Palmer's swimmers	*sw*	*sw sw*	without special care: soon after birth	9
Myelopathy	*mp*	*mp mp*	4–9 months	9
Bjerkås' leukodystrophy	*ld*	*ld ld*	3–6 months	9
Deficiency of factor X	*H10*	*H10 H10*	in first 10 days	10
von Willebrand's disease	*Vw*	*Vw Vw*	stillborn	10
Cyclic neutropenia (gray Collie)	*cn*	*cn cn*	for most, under 6 months	10
Renal cortical hypoplasia	—	many	within 2 years	12
Neuronal ceroid-lipofuscinosis	*li*	all *li li*	within 2 years	12

[a]These symbols are suggested because none had been previously assigned. Capital letters indicate dominant genes; lowercase letters, recessive ones. No symbols are assigned to three characters for which the genetic basis is unproven. Symbols suggested in the future should differ from any preempted by earlier workers; for example, from Little's genes for color (Table 8-2).

Appendix II

A Letter to Veterinarians

Dear veterinarians:

As its title tells you, this book was not written for you—but for your clients. There is much less in it about pathology than about genetics, but I have given (briefly) clinical symptoms by which the genetic defects may be recognized. I have also tried insofar as possible to use terms that the dog breeders already know. Front legs are front legs—and not thoracic limbs. I hope that some readers will pester you a bit for more information about the pathology caused by some of the genetic defects. That is your field—not mine.

Defects for which genetic evidence is scarcely good enough have been left out, and there are probably others of which I should know, but don't. I suggest that you check also in Patterson's "Catalogue of Genetic Disorders of the Dog" (1977). It has extensive references.

I hope that some of you who read this were among the veterinary students at Cornell to whom I taught animal genetics during a period of 20 years. From the number of them who have since written to me about specific problems, there is some reason to believe that the course (which was the first of its kind in American veterinary colleges) may not have been a case of love's labor lost. Even though one of you confessed (years later) that what he remembered best of that course were a few lines of Shakespeare's *Julius Caesar* that I had once innocently tossed in to enliven an otherwise dull period, I like to feel that the bit of genetics you got did help in your later years.

My purpose in writing this note is to encourage you to help in advancing our knowledge of hereditary defects in dogs. You are more likely than any geneticist to find abnormalities that seem to be unduly frequent in some breed, in some kennels, perhaps even in some litters. Please don't try to fit them all into simple 3 : 1 ratios. Some of them (like hip dysplasia and the Collie eye anomaly) are more likely to be polygenic.

If your time and facilities don't permit you to make the experimental matings necessary to find the mode of inheritance, perhaps (as several veterinarians have done) you can persuade the dog breeders to do it in the interests of science, and of the breed in which the defect occurs. As an example of what even one more pup can contribute to science, see how much one concerned dog breeder and one female pup were able to add to our knowledge of the globoid-cell leukodystrophy studied by Hirth and Nielsen (Chapter 9).

If all else fails, why not pass the word about any previously unknown abnormality to some of your confrères who can study it? Some of them specialize in hereditary defects of the blood; others are interested in anatomical defects; some are specialists in diseases of the nervous system; others specialize in abnormalities of metabolism. Most of these are in veterinary colleges or state laboratories and can enlist the help of geneticists to set up experimental matings and to interpret results.

If you do undertake a genetic study, perhaps using records from cooperating breeders, please don't confine your data to the numbers of *affected* animals in each litter—be sure to include the numbers of *normal* litter-mates, and also, the sex of every pup. Try to get a classification for every pup in each litter.

If you publish a pedigree of the condition, it will help your readers if you use these standard symbols:

□	○	■	●
Male, normal	Female, normal	Male, affected	Female, affected

You can put a dot in the center (⊡ ⊙) for proven carriers, and add your own improvisations for dogs not classified. See Figure 12-3.

Please help. Good luck!

Frederick B. Hutt

P.S. Be sure to bone up on those congenital abnormalities of the heart (Table 10-6) before your clients start asking about them.

Appendix III

Laboratory Resources for Genetic Defects

At the institutions listed below, there are (in 1978) laboratories or departments specially interested in genetic defects in dogs. Breeders will find in them veterinarians willing to investigate defects that appear to be hereditary but have not yet been thoroughly studied.

Names are given of only a few key people, but these have collaborating associates.

MICHIGAN STATE UNIVERSITY, EAST LANSING, MI 48824

Department of Medicine. Dr. R. W. Bull and associates. Special interest in blood antigens and blood typing. Will make tests for paternity (for which a fee is charged).

Department of Pathology. Dr. G. A. Padgett. Interests include all genetic defects.

NEW YORK STATE DEPARTMENT OF HEALTH

Division of Laboratories and Research, Tower Building, Empire State Plaza, Albany, NY 12237. Dr. W. Jean Dodds. Special interests in blood, defects of blood clotting, and disorders related thereto.

UNIVERSITY OF PENNSYLVANIA, SCHOOL OF VETERINARY MEDICINE, PHILADELPHIA, PA 19104.

Section of Medical Genetics. Dr. D. F. Patterson, Director, and associates. Special interest in congenital defects of the heart, but also in various other genetic abnormalities in dogs.

Department of Ophthalmology. Drs. L. F. Rubin and G. Aguirre. Hereditary abnormalities of the eyes.

WASHINGTON STATE UNIVERSITY, PULLMAN, WA 99164

Department of Veterinary Microbiology and Pathology. Drs. G. A. Hegreberg, J. R. Gorham, and associates. Interests include all genetic defects in dogs.

Index of Breeds Mentioned

Index of Topics